AF615644

# Microelectrofluidic Systems

## Modeling and Simulation

## *Nano- and Microscience, Engineering, Technology, and Medicine Series*

*Series Editor*
Sergey Edward Lyshevski

*Titles in the Series*

**MEMS and NEMS: Systems, Devices, and Structures**
Sergey Edward Lyshevski

**Microelectrofluidic Systems: Modeling and Simulation**
Tianhao Zhang, Krishnendu Chakrabarty, and Richard B. Fair

**Nano- and Micro-Electromechanical Systems: Fundamentals of Nano- and Microengineering**
Sergey Edward Lyshevski

*Forthcoming*

**Nanodynamics in Engineering and Biology**
Michael Pycraft Hughes

# Microelectrofluidic Systems

## Modeling and Simulation

Tianhao Zhang
Cadence Design Systems Inc., Cary, North Carolina

Krishnendu Chakrabarty
Duke University, Durham, North Carolina

Richard B. Fair
Duke University, Durham, North Carolina

CRC PRESS

Boca Raton London New York Washington, D.C.

**Library of Congress Cataloging-in-Publication Data**

Zhang, Tianhao, 1970-
Microelectrofluidic systems : modeling and simulation / Tianhao Zhang, Krishnendu Chakrabarty, Richard B. Fair.
p. ; cm. — (Nano- and microscience, engineering, technology, and medicine series)
Includes bibliographical references and index.
ISBN 0-8493-1276-0 (alk. paper)
1. Biomedical engineering. 2. Microelectromechanical systems. 3. Nanotechnology. 4. Polymerase chain reaction—Automation. I. Chakrabarty, Krishnendu. II. Fair, Richard B. III. Title. IV. Series.
[DNLM: 1. Models, Chemical. 2. Biomedical Engineering—methods. 3. Microchemistry. 4. Polymerase Chain Reaction. QU 25 Z63m 2002]
R857.N34 Z48 2002
610'.28—dc21 2002019344



Direct all inquiries to CRC Press LLC, 2000 N.W. Corporate Blvd., Boca Raton, Florida 33431.



**Visit the CRC Press Web site at www.crcpress.com**


International Standard Book Number 0-8493-1276-0
Library of Congress Card Number 2002019344
Printed in the United States of America 1 2 3 4 5 6 7 8 9 0
Printed on acid-free paper

# *Acknowledgments*

We are grateful to Nora Konopka of CRC Press for encouraging us to pursue this book project. We are also grateful to IEEE, Elsevier Science, Applied Computational Research Society for granting us copyright permission to use materials from our published work. This book grew out of a research project supported by the Defense Advanced Research Projects Agency (DARPA). We thank DARPA for supporting this work through the MTO Composite CAD program. In particular, the authors would like to thank Dr. Anantha Krishnan at MTO in DARPA and Mr. Robert Hillman at the Air Force Research Laboratory for supporting this work. The authors also acknowledge the work of the late Dr. Allen M. Dewey who initiated the research effort at Duke University in microfluidic systems design. Finally, we acknowledge the contribution of Jie Ding, Hong Ren, Feng Cao, Vijay Srinivasan, Jason A. Jopling, and numerous other colleagues who contributed to this research project.

# *Contents*

# Preface

This book is about integrated hierarchical modeling, simulation, and design of microelectrofluidic systems (MEFS). The goal of the book is to position top-down modeling and simulation in the context of MEFS design, as well as to generate interest and motivate research on this important topic.

Composite microsystems integrating microelectromechanical and microelectrofluidic components with electronics are emerging as the next generation of system-on-a-chip (SOC) designs. MEFS increasingly are being used for automated drug dispensing and micro-chemical analysis (DNA analysis and lab-on-a-chip). Nevertheless, there remain several roadblocks to rapid and efficient composite system integration. Primary among these is the need for modeling techniques, and simulation and optimization tools. This book responds to a pressing need for a structured methodology for MEFS design automation. In this book, the reader will find in one integrated volume, top-down design automation approaches for MEFS. Readers with a background in electronic design automation will use this book to apply their expertise to composite system design. On the other hand, readers from the fluidics domain, who have thus far been developing bottom-up inflexible and custom microsystems, will find this book invaluable in adopting a more top-down and generic design methodology.

The book is based on the premise that top-down modeling, simulation and optimization offer promising solutions to the problems encountered in MEFS design. It includes MEFS hierarchical modeling, a hierarchical design environment, performance evaluation, and hierarchical optimization. A definition of basic variables and elements needed to describe MEFS behavior is first presented. MEFS behavior is modeled across three layers of abstraction: low-level component layer, high-level reconfigurable architecture layer, and bio/chemical application layer. In addition, the suitability of several existing simulation languages is evaluated for hierarchical design, and a hierarchical integrated design environment with SystemC is developed. Its architecture and associated functional packages are presented.

MEFS performance analysis is difficult because coupled-energy behavior creates strong links between high-level architecture and low-level component design parameter variations. This problem is addressed by trading-off behavioral fidelity with analysis efficiency to develop a hierarchical modeling and simulation methodology. This methodology encompasses both architectural system simulation with stochastic macromodeling and component simulation with lumped-element nodal modeling.

Using the integrated design environment based on SystemC, this methodology is evaluated by applying it to a micro-chemical handling system.

Due to growing design complexity, fabrication process variations, and the harsh operating environments of MEFS, there is a need for hierarchical design optimization to support all aspects of product development. Several MEFS design optimization methodologies are demonstrated in this book. They include a statistical response analysis strategy for on-target design and process optimization, a robust design methodology, and a new application flexibility design methodology to leverage the hardware/software co-design principle. A number of representative MEFS devices are designed to illustrate these optimization algorithms. Finally, a performance comparison is presented between two types of microfluidic systems—continuous-flow systems and droplet-based systems. The comparison is based on a specific microfluidic application—a polymerase chain reaction (PCR) system. The modeling and simulation of PCR are based on the SystemC design environment. The comparison includes throughput, processing capacity, correction capacity, and design complexity.

To the best of our knowledge, this is the first book that presents a comprehensive integrated MEFS design strategy. It combines top-down and bottom-up design philosophies, and it supports hierarchical modeling and simulation from the component level to the system level. In addition, it leads to multi-objective optimization tools that address design tasks from conceptualization to final manufacturing. Moreover, by using a unique modeling and simulation language—SystemC, this design approach potentially leads to a decrease in design time and life-cycle maintenance costs.

This book grew out of an ongoing research project on composite microsystems at Duke University. The results of this research have been published as papers in a number of journals and conference proceedings. The chapters in this book present all these results as a research monograph in a single volume. It can be used as a textbook for a graduate course, with a course title such as "Design Automation of Composite Microsystems." While the book is primarily directed at researchers and graduate students in electrical and biomedical engineering, it will also be useful as a reference for academic and industrial researchers in microelectrofluidics and electronic design automation.

In summary, this book is expected to pave the way for integrated top-down design of hierarchical MEFS. The framework described in this book will reduce design time and design cost, and increase system reliability.

# List of Figures

# List of Tables

# Chapter 1

# Introduction

Composite microsystems that incorporate microelectromechanical and microelectrofluidic devices are emerging as the next generation of system-on-a-chip (SOC) designs. These systems combine microstructures with solid-state electronics to integrate multiple energy domains. The combination of microelectronics and microstructures, and the integration of electrical, mechanical, and fluidic domains enable new classes of integrated systems targeted at environmental sensing, control actuation, biomedical analyses, agent detection, and precision fluid dispensing [2, 10, 11].

Microelectrofluidic systems (MEFS) is an area of research that addresses the miniaturization of composite devices and systems, and the study of new applications associated with the handling of liquids and gases. Microfluidics not only offers the obvious advantage of size reduction (small medical implants [12] and minimally invasive surgery [13]), but it also reduces power dissipation and increases system reliability. Microfluidics offers unique new possibilities in controlling small amounts of fluids for precision dispensing (micro dosing [14]), and reducing reagent consumption for on-line chemical analysis and real-time process monitoring. By scaling down the concentrations of chemical samples, simpler sensing techniques can be utilized to replace present, more costly, practices. These practices involve batch analysis, sample pre-treatment, and frequent calibration. Smaller sample volumes reduce storage costs, facilitate uninterrupted use, and benefit medical procedures in numerous ways [10].

Significant progress has been reported recently in the design of individual MEFS components. For example, switches [15], valves [16], channels [17], pumps [18], etc. have been designed, built, and studied extensively. A fluidic mixer and micro glass channels, designed and manufactured by Kymata Netherlands, Livingston, Scotland, are shown in Figure 1.1 and 1.2, respectively [1]. Small systems that combine existing components into useful devices have also been designed and built. For instance, Figure 1.3 shows the prototype of a DNA analysis device built at the University of Michigan [2], and Figure 1.4 shows a liquid analysis system developed by Sandia National Laboratories [3].

## 1.1 Modeling and Simulation Issues

It is useful to identify analogies between microelectronics and microfluidics in order to position microfluidics on a maturation path that best exploits microelectronics technology. Just as progress in component-level digital logic design enabled and motivated the development of computer architecture, similar developments in component-level microfluidic devices are enabling and motivating the microfluidic molecular systems (microflumes) architecture. Though significant progress has been

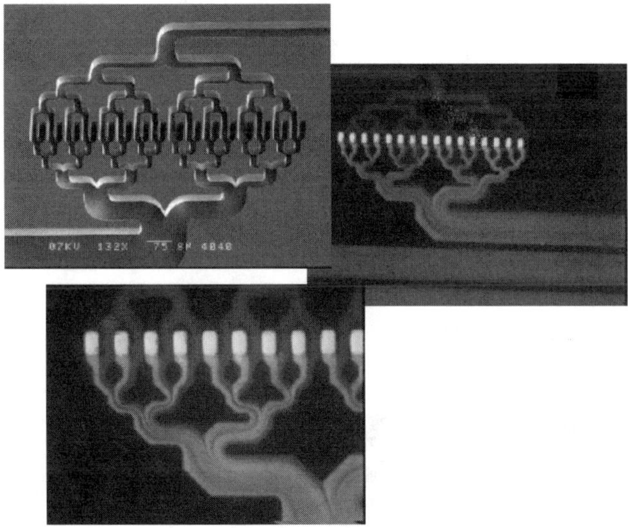

**FIGURE 1.1**
**An example of microelectrofluidic systems: A micromixer (Kymata Netherlands, Livingston, Scotland [1]).**

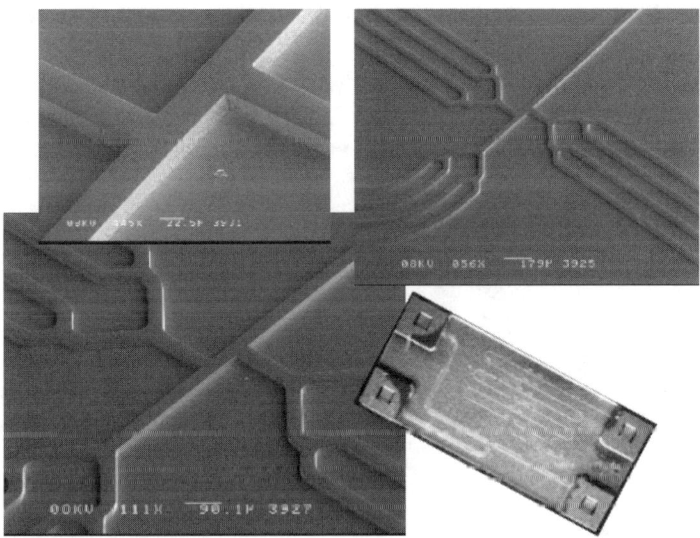

*SEMs and photo of the micro glass channels*

**FIGURE 1.2**
**An example of microelectrofluidic systems: Micro glass channels (Kymata Netherlands, Livingston, Scotland [1]).**

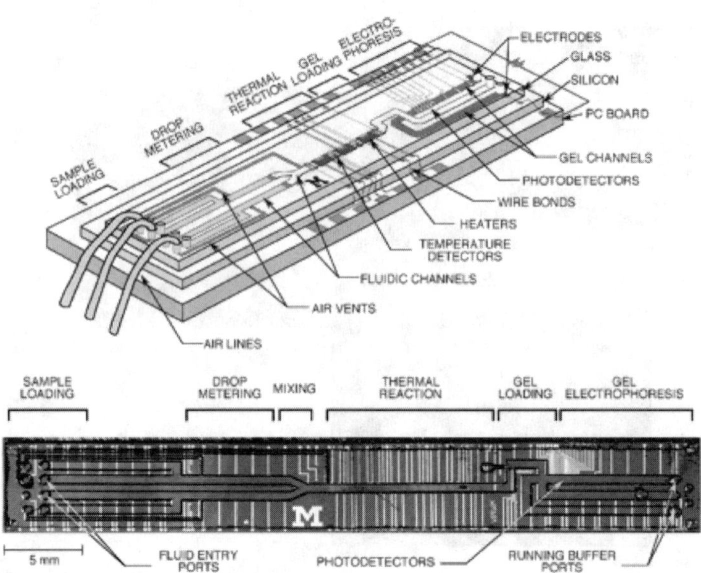


**FIGURE 1.3**
**An example of microelectrofluidic systems: DNA analysis device (University of Michigan [2]). (Reprinted with permission from M. A. Burns *et al*, An integrated nanoliter DNA analysis device, *Science*, vol. 282, pp. 484-487, 1998. Copyright 2002 American Association for the Advancement of Science.)**

**FIGURE 1.4**
**An example of microelectrofluidic systems: A user injects a sample into the liquid analysis system of the hand-held, integrated device for analyzing liquid and gas mixtures (under development at Sandia National Laboratories [3]).**

made in exploring elemental microfluidic devices, little progress has been made in understanding how the capability of these devices can be exploited and utilized to address system applications involving fluid acquisition, storage, transport, reactions, and dispensing. Examples of microfluidic systems built to date involve either highly specialized applications, such as immunobiosensing, or more general-purpose applications addressing primarily only one aspect of fluidics processing, such as transport. Thus, to establish a maturation growth path (technology and commercialization) for microfluidics similar to the maturation growth path for microelectronics, there is a need to define basic architectural organization and execution concepts for assembling various microfluidic devices into a network. This network can perform a variety of tasks supporting a diverse set of applications. To address the need to define basic architectural organization and execution concepts for microelectrofluidics, as shown in Figure 1.5, new architectural concepts must be developed across the following three layers of abstraction. These layers integrate the state-of-the-art microfluidic components into a more encompassing microliquid handling system. This system can be readily reconfigured and reused to enact a variety of biomedical chemical detection, analysis, diagnostic, and dispensing applications.

1. Biomedical/Chemical Application Layer
2. Reconfigurable Microliquid Handling Architecture Layer
3. Microfluidic Component Layer

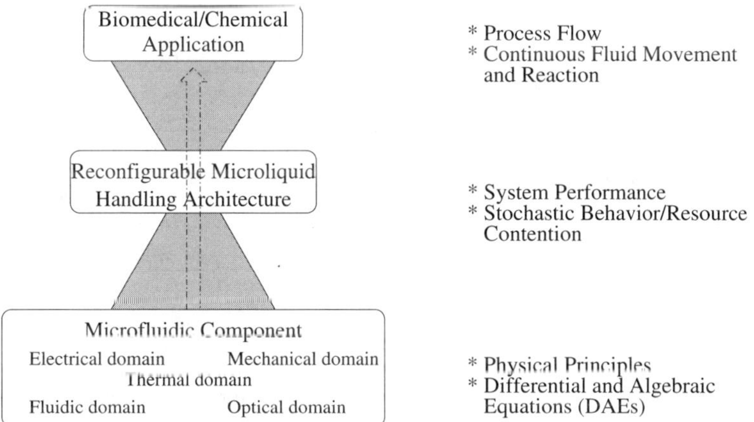


**FIGURE 1.5**
**MEFS integrated hierarchical system perspective includes three levels of abstraction. Each level of abstraction presents unique model fidelity, domain representation, and simulation efficiency requirements.**

In addition, system complexity—measured by the growing number of devices and the increasing levels of heterogeneity due to the new coupled-energy or composite designs—necessitates a scalable design method to address both the application complexity and the component capacity. The application complexity is a measure of how the performance of the reconfigurable microliquid handling system architecture scales with increasingly complex chemical and biological analyses. The component capacity investigates how the performance of the reconfigurable microliquid handling system scales with advances in constituent microfluidic device technology.

In order to facilitate scalable design and the subsequent evolution of MEFS, more extensive system modeling and simulation capabilities are required, as shown in Figure 1.5. Each layer of abstraction presents challenges involving model fidelity, domain representation, and simulation efficiency. The biomedical application layer is the highest level of abstraction. It involves process flow modeling, as well as the simulation of continuous fluid movement and chemical/biological reactions directed towards an application, such as microdialysis, chemotherapy, genetic analysis, or cell filtration. The evaluation of biomedical applications requires lower-level information, such as channel pressure drops and pump throughput, which are associated with the second level of abstraction—the reconfigurable microliquid handling architecture. This level of abstraction involves performance modeling and simulation of the stochastic behavior of the major component/resources and their aggregate operation in executing a biomedical application. Evaluating the architectural performance involves, in turn, lower-level information, such as microfluidic transport and the device operation, which are associated with the lowest level of abstraction—microfluidic components. This level of abstraction involves detailed circuit simulation requiring integro-differential and algebraic equations, which characterize physical properties and processes. Here, the microfluidic component layer of abstraction is defined as the circuit level, while the biomedical/chemical application and reconfigurable microelectrofluidic system architecture are defined as the system level.

## 1.2 Modeling and Simulation Needs

While MEMS design tools have reached a certain level of maturity, MEFS CAD is still in its infancy. Since the concept of special CAD systems for MEMS was first proposed at Transducer'87 [19], several research groups have reported significant progress in this area, and some commercial microsystem CAD tools have now been developed [20, 21, 22, 23, 24, 25]. The MEMCAD system from Coventor, Inc., the IntelliSuite$^{TM}$ system from IntelliSence Corporation, the VULCAIN$^{TM}$ generic MEMS engineering design platform from MEMSCAP, and the ANSYS system from ANSYS, Inc., are focusing on MEMS design. Many commercial CFD (Computational Fluid Dynamics) tools, such as CFD-ACE+ from CFD Research Corporation and FlumeCAD from Coventor, Inc. are also available. These tools mainly focus on the thermal and structure modeling and design, thus new solutions are necessary for the design of current and next generation MEFS.

### 1.2.1 Computational Architectures for MEFS

The limitation of present designs is that devices tend to be application/analysis specific. There is no one system capable of performing a collection of differing analyses or procedures. In addition, in a number of systems, only the analysis device is microfluidic in nature. The remaining pre-/postprocessing is performed by normal macroscopic, possibly automated, methods. It is easy to recognize the potential of a universal architecture, one in which multiple, differing procedures may be performed with simple reconfiguration, and possibly in parallel. Current MEMS CAD techniques are not focused on such reconfigurable and reusable computational MEF architectures, the design of such an architecture, and simulation and performance modeling.

### 1.2.2 Hierarchical Modeling and Simulation

The close integration of devices is associated with strong energy-coupling issues. In order to support the growing complexity of MEFS design and to carry out global performance optimization, system-level performance analysis methodologies and tools are needed. These methods and tools must not only incorporate phenomenological laws from multiple disciplines, but they must also characterize dynamical behavior ranging from overall application execution to individual component operation. Moreover, the levels of abstraction need to be linked to correlate the multiple analyses in both the top-down design and the bottom-up verification. In other words, the multiple levels of abstractions/analyses need to be hierarchically integrated. Existing MEMS CAD tools and the underlying hierarchical modeling and simulation techniques lack hierarchical modeling and simulation capabilities for MEFS design. They mainly focus on low-level components, and pay insufficient attention to analysis methodologies that are necessary to understand how the capabilities of these devices can be exploited and utilized at the system level. These capabilities are needed to represent the system design perspective of composite microsystems from functional unit level to system performance level.

### 1.2.3 Advanced Hierarchical Design Methodology

MEFS design methods should exploit state-of-the-art reconfigurable SOC system design techniques. These techniques rely on a combination of top-down decomposition and bottom-up aggregation. Moreover, hardware/software co-design methodologies are necessary, which emphasize system functional unit reusability to achieve short design cycle time. These are the key methods for significantly pushing performance envelopes for a wider range of applications. However, the current MEMS CAD tools only emphasize the component oriented bottom-up design methodology, and the popularly used hardware description languages for microfluidic device modeling limit the advanced top-down system design.

### 1.2.4 Hierarchical Design and Simulation Optimization

As the number of pilot applications of integrated MEFS grows, there is a need for design optimization to support all aspects of product development, including design, manufacturing, and operational use from the component level to the system level. Current design optimization strategies focus on improving manufacturing yield via time-consuming statistical models. They lack robust optimization and hierarchical verification capabilities for different optimization objectives. In addition, the single-component-oriented design perspective prevents the extension of the optimization methods to hierarchical multiple-device system-level optimization.

### 1.2.5 System Design Language Uniformity

Traditionally, different modeling and simulation languages are used for describing the unique set of representational conventions and simulation methodologies for each level of hierarchy. The logic modeling languages, for instance VHDL/VHDL-AMS and Verilog, are used for the low-level functional unit design [18]. For the higher-level of abstractions, e.g. biomedical application level and microelectrofluidic system architecture level, process-interaction and continuous-perspective-oriented modeling and simulation languages have been used. These include SIMSCRIPT II.5 [26], SLAM [27], and general-purpose software programming languages, such as C and Ada. However, this system design approach requires extensive human interaction. It also leads to problems of misinterpretation of concept specifications in the translation between different data models and tools. Thus, it is beneficial to use a modeling and simulation environment using a common language and a coordinated set of simulation engines to support each level of abstraction. The potential benefits of this approach include improvements in design time and life-cycle maintenance costs.

## 1.3 Overview

The Gasjki and Kuhn's Y-chart [28] has long been used as a conceptual framework for VLSI design. It has three domains of design description: Behavioral, Structural and Physical. Each domain has three levels of abstraction, individually. The design process is represented by step-wise refinement in all the three domains from outer levels towards the center. On the analogy of this Y-chart in microelectronics CAD, we present a MEFS CAD closed-loop integration strategy, as shown in Figure 1.6. It includes three domains for system design: the integrated microsystem conceptualization, modeling and simulation; the microsystem design optimization; and the microsystem validation and fabrication. In addition, the system design is extended from component level to system level. All of these design domains have their unique characteristics, but they are also tightly coupled to each other. The designer first transfers the design idea into a model, which captures the most important properties of a microsystem and provides a good behavioral approximation that shortens simulation time. The optimization task is to design the system to match the different optimization objectives. The goal of the validation task is to verify the optimal design results, so that they can be used for further product fabrication. The system model and associated system simulation provide useful data for design optimization. A high degree of system model accuracy and simulation efficiency are required in this task. System optimization not only considers the performance objective, but it is also concerned with the fabrication environment so that the manufactured product can match the design requirement. With a product sample, the accuracy of the models and the validation of optimization results can be tested. The MEFS design environment must support this closed-loop integration.

Thus, as shown in Figure 1.7, the goal of this book is to present a new approach for integrated modeling and simulation of MEFS. To the best of our knowledge, this is the first attempt to develop a comprehensive integrated MEFS design environment. This new approach includes four aspects: hierarchical modeling, hierarchical integrated design environment, hierarchical simulation and performance evaluation, and hierarchical optimization. This perspective on MEFS design combines top-down and bottom-up design philosophies. It also supports hierarchical modeling and simulation from the component level to the system level. In addition, it leads to multi-objective optimization tools that address design tasks from conceptualization to final manufacturing. Moreover, by using a single modeling and simulation language, it is possible to make improvement in design time.

The contributions of this book include the following:

1. A definition of the basic variables and elements needed to describe MEFS behavior from the low component level to the architectural and biomedical/chemical

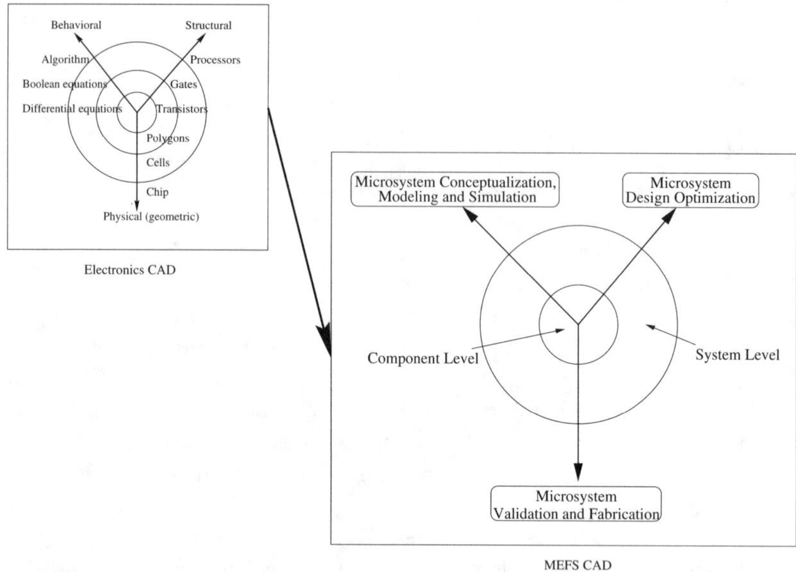


**FIGURE 1.6**
**System perspective of the composite microsystem closed-loop design integration.**

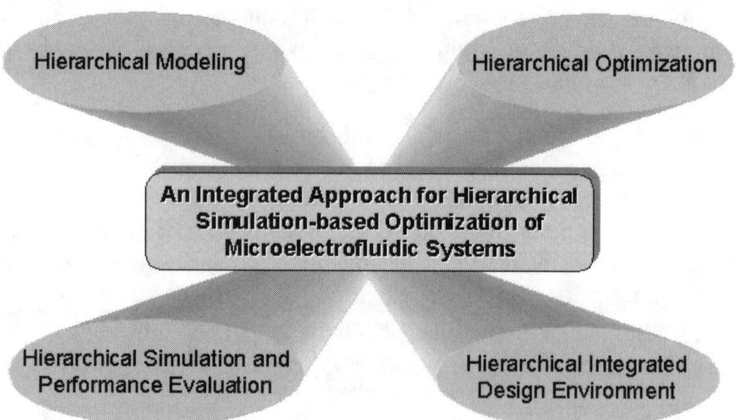


**FIGURE 1.7**
**Research content consists of hierarchical modeling, hierarchical simulation and performance evaluation, hierarchical optimization, and hierarchical integrated design environment.**

application levels.

2. A comparison of the suitability of several simulation languages for hierarchical design. A hierarchical integrated design environment with SystemC. This environment models the system at each level without losing information.

3. MEFS hierarchical modeling and simulation methodologies that are tailored to the design requirements and the system performance goals. Additionally, the application of this methodology is illustrated for a special MEF system with SystemC design environment.

4. An on-target statistical approach that finds the design solution more efficiently. As a case study, a microvalve modeled with SystemC is presented.

5. A robust design methodology to reduce device performance sensitivity on the design parameter variations. An electrostatic-comb microresonator, as a representative of MEMS, and a microvalve are used to illustrate this methodology.

6. An application-flexibility design methodology that extends the use of MEFS to more application areas. This methodology makes devices match the customer's different design requirements, and helps address the problem of low manufacturing volume. A microvalve is used to illustrate this methodology.

7. A more general microelectrofluidic system computational architecture involving a multi-drop bus, and a pipelined structure for continuous-flow systems. Functional requirements are explained, and results of performance modeling and analysis of the microfluidic processor architecture are presented.

8. A performance comparison of two types of microfluidic systems continuous flow systems and droplet-based systems—based on the SystemC design environment. The comparison is based on a special microfluidic application—a polymerase chain reaction (PCR) system. The comparison metrics includes the system throughput, processing capacity, correction capacity, and design complexity.

9. An architectural design and optimization methodology for performing biochemical reactions using two-dimensional electrowetting arrays. As a case study, the optimization method is applied to a polymerase chain reaction (PCR) system.

In the following chapters, each of these contributions is described in more detail. The organization of the book is as follows.

A MEFS hierarchical modeling strategy is presented in Chapter 2. A circuit-level and system-level performance modeling and simulation strategy is developed. These strategies are useful to represent the MEFS component behavior, the reconfigurable architecture and biomedical/chemical applications in detail.

Next, depending on the generic requirements of MEFS hierarchical modeling and simulation, an integrated MEFS design environment is developed in Chapter 3. At first, the suitability of several existing simulation languages is evaluated for hierarchical design problems. These languages include VHDL/VHDL-AMS, Matlab, C/C++, SLAM, and SystemC. Then, an approach is presented that uses SystemC as a potential candidate to build a complete system modeling and simulation environment. This environment represents system behavior at each level without losing information. In addition, based on the proposed MEFS CAD strategy, the architecture and the associated functional packages of this unique hierarchical integrated design environment are discussed.

Hierarchical modeling and simulation methodologies are presented in Chapter 4. These methodologies combine a high-level stochastic queuing network approach with low-level nodal conservative differential equations. A more general microelectrofluidic system computational architecture for continuous-flow systems is introduced, which involve a multi-drop bus and pipelined structure. Based on this generic reconfigurable architecture, a MEFS reconfigurable architecture optimization strategy is developed that addresses system performance bottlenecks. The system is designed to operate in the saturation mode using the traffic variation method [29]. Using the SystemC integrated design environment, these methodologies are evaluated by applying them to a special MEFS case—a micro-chemical handling system.

In addition to MEFS modeling and simulation methodologies, novel simulation-based design and process optimization algorithms are also parts of the integrated design framework. These algorithms are described in Chapter 5. At first, two computationally efficient simulation-based design strategies are discussed. These methodologies have been used to provide simulation data for the design optimization algorithms. A validation strategy is presented to verify the optimization algorithms. To make the design solution match the design performance requirement efficiently, an optimal on-target design algorithm was presented. After meeting the system performance requirement, the second task for MEFS optimization involves operation reliability and design robustness. By leveraging several advanced optimization methods, a MEFS optimization methodology is proposed based on the Taguchi robust design method [30] and the statistical response analysis method [31]. In addition, to extend a MEFS design to a wider application area, a novel design approach for application-flexibility is proposed using the analogy to hardware/software co-design methodology [32]. Special MEFS and MEMS devices are used as case studies to illustrate these optimization algorithms.

A performance comparison between two types of microfluidic systems—continuous-flow systems and droplet-based systems—is presented in Chapter 6. The comparison is based on a special microfluidic application—a polymerase chain reaction (PCR) system. The modeling and simulation of PCR are based on the SystemC design environment. Physical implementation and the system performance issues are discussed. It is shown that the droplet-based microfluidic system possesses higher performance capacity, as well as lower design and integration complexity. Then, an architectural design and optimization methodology is presented for performing droplet-based biochemical reactions using two-dimensional electrowetting arrays. Integer linear programming is used to perform optimization objectives. As a case study, the optimization method is applied to a polymerase chain reaction (PCR) system.

Finally, in Chapter 7, we present conclusions and outline future direction for MEFS CAD.

The appendices contain useful information about the VHDL-AMS and SystemC software that were developed to evaluate the proposed methodologies.

# Chapter 2

# Hierarchical Modeling

MEFS complexity arises due to the growing number of devices and the increasing levels of heterogeneous coupled-energy domains. As a result, system design is evolving into a multidisciplinary field. This broad design scope requires a comprehensive model to study the dynamic behavior of the system. This model should capture the most important properties of the microsystem, and provide an accurate behavioral approximation. Although it is theoretically possible to build a single complicated model to describe the MEFS behavior, the overall system is still too complex to be handled as one entity. Thus, modeling MEFS behavior consists of two integral parts: system-level modeling and component modeling.

System-level modeling involves performance modeling and behavioral simulation for specific biomedical and chemical applications. For a chemical handling system, system modeling primarily considers the system throughput and latency. It neglects detailed implementation issues. In contrast, component modeling investigates the individual microfluidic component behavior, and emphasizes the definition of physical properties and relationships at the circuit level. Component modeling therefore offers an approach that is complementary to system-level simulation.

In this chapter, fundamental variables and elements needed to describe MEFS characteristics are defined from the lower component level to the higher system level. The basic variables for the circuit-level dynamic behavior of multiple energy domains are presented in Section 2.1. Then, based on the characteristics of MEFS architecture and the stochastic nature of an application's behavior, the fundamental variables are defined for these system-level features in Section 2.2. These fundamental variables capture the MEFS behavior, and they are critical requirements for the MEFS modeling and simulation language. A summary of MEFS modeling issues is presented in Section 2.3.

## 2.1 MEFS Dynamic Modeling and Simulation at Circuit Level

Direct numerical simulation of the dynamic behavior of MEFS devices requires three-dimensional, distributed nonlinear models. These models are computationally difficult, and very expensive to use. Therefore, as illustrated in Figure 2.1, the behavior macro-model needs to be built.

- Direct numerical dynamical simulation of fully-meshed distributed nonlinear devices is computationally difficult, and very expensive. Device level energy-coupling modeling and simulation are done in the *quasi-state* model, which does not give information about dynamic behavior characterizations.
- In order to study the system dynamic behavior, it is necessary to reduce the

number of degrees of freedom, from the meshed device three-dimensional models to two-dimensional lumped-element models, which presents the overall system behavior as a function of time.

- In addition to 3D component level energy-coupling modeling and simulation, building the 2D circuit level behavior macro-model is necessary to obtain the global optimal solution.
- Moreover, the use of marco-model of microsystem devices results in compact and efficient representations, and it avoids convergence problems due to the coupling of heterogeneous simulators.

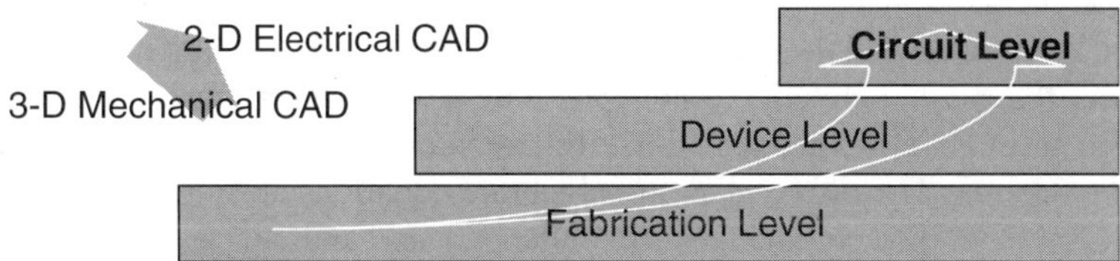


**FIGURE 2.1**
**Abstracting composite microsystem dynamical models to higher levels of abstraction**

The circuit level representation abstracts three-dimensional, distributed effects into a two-dimensional network of lumped parameter elements governed by a system of ordinary differential and algebraic equations (ODAEs). To establish the microsystem dynamic modeling and simulation at the circuit level, several research questions need to be addressed:

- *Is it appropriate to describe MEFS dynamic behavior with lumped-element model by using ODAEs?*
  Although it is necessary to abstract the three-dimensional model to the two-dimensional network of lumped parameter model by ODAEs, a study is needed to understand this abstraction's possibility, accuracy and limitation.
- *What are the fundamental variables to simulate the multiple-energy domain behavior?*
  Different energy domain components possess various fundamental elements to describe their unique characteristics. Hence it is necessary to recognize the fundamental variables to describe system dynamic behavior in multiple-energy domains: electrical, mechanical, fluidic and thermal. Additionally, coupled-energy domain characteristics show that energy is always transferred from one energy domain to another. It is necessary to understand the modeling and simulation requirement for describing the transduction behavior.

- *What methodology should be used to represent the relationships between these fundamental elements?*
  After the definition of system variables, the next step is to formulate a dynamic model of the physical system in the form of mathematical relationships. Depending on the microsystem coupled-energy domain situation, energy-law based methods are used to study the conservative relations between different element components.

In Section 2.1.1, the fundamental physical principles for different energy domains are discussed, and it is shown that the lumped-element models with ordinary differential and algebraic equations (ODAEs) can provide a common characterization spanning multiple energy domains. In addition, in Section 2.1.2, the fundamental variables of multiple energy domains are described. Their constitutive relations are presented in Section 2.1.3. Based on these fundamental variables and their relations, Section 2.1.4 discusses the circuit network topology defining the global structure of the equations, and also presents element constitutive relations defining specific terms within the global structure. The differential and algebraic equations are formed within a systematic framework recognizing linear independence and energy conservation. In Section 2.1.5, an equivalent circuit approach for MEFS circuit-level modeling and simulation is presented.

### 2.1.1 Classification of Dynamic System Models

Mathematical models are needed to study the dynamic behavior of composite microsystems. These models can always be classified into two categories based on the nature of the underlying differential equations [33]:

- Distributed-element models
  When the dynamic behavior of the system requires more than one independent variable to describe, partial differential and algebraic equations (PDAEs) are used. These types of mathematical models are called "field" models or *distributed-element* models. Distributed-element models with PDAEs allow "complete freedom" in the description of dynamic systems. They behave almost exactly like the real systems.

- Lumped-element models
  In order to reduce the complexity of the mathematical expressions, and reduce the time needed to obtain numerical solutions for distributed-element models, simpler lumped-element models with ODAEs are often used. In contrast to the characteristics of distributed-element models, lumped-element models concentrate matter and energy into discrete "lumps", and the variables at a given spatial location in each lump are used to represent the variables of other regions in this lump. The number of lumps and the lump size must be decided

when lumped-element models are used. With more lumps, lumped-element models can become more accurate and closer to the distributed-element model.

A wavelength/physical size concept can be used to explain the rationale of building lumped-element models for any physical system that exhibits wave propagation, such as electromagnetic systems, mechanical vibrating systems and acoustic systems [33]. The wavelength/physical size criterion is shown in (2.1). The key concept here is that if the physical size of a device is small compared to the wavelength associated with signal propagation, the device may be considered lumped, and a network lumped-element model can be employed.

$$Wavelength\ \lambda = \frac{velocity\ V\ \ of\ \ wave\ \ propagation}{signal\ \ frequency\ f} \tag{2.1}$$

The propagation velocity of electrical waves in free space is 300000 kilometers per second, and assuming a representative value of 200MHz for the system operating frequency, the wavelength of the system is 1.5 meters per cycle. Typically, electrical components are much smaller than 1.5 meters, thus microelectrical systems can be treated with the simple lumped-element approach.

In addition, the essential solid body characteristics allow micromechanical systems to be analyzed as lumped-element models [34]. The variables at a given spatial location in a solid body can always be used to represent the variables of other regions. In order to account for the elasticity and tortures of deformation, the solid body has to be cut into several lumps. Then, by applying a pertinent physical law to the solid body, a set of simultaneous ODAEs are generated. The solution of these equations describe the system dynamic behavior.

However, in contrast to the electrical and mechanical energy domains, matter and energy may not be continuously distributed over the space within some fluidic systems. In addition, due to the generally less well-defined shapes of bodies of fluid (as compared to solid bodies), microfluidic systems appear to be less suited for the lumped-element viewpoint. Since every spatial location has its own flow rate and direction of flow, using a given spatial point to be representative of the local environment may cause behavioral description errors. Nevertheless, using a lumped element model to describe microfluidic systems is appropriate when the fluidic flow is laminar, the fluid is incompressible, and the fluid shape is well defined [33]. For instance, when a fluidic sample flows in a channel whose diameter is very small ($mm$), the fluidic flow can be described using the lumped-element model. Within a given element there is no variation, but behavior such as pressure and velocity, usually change between different elements. It is clear that when a model is made up of a number of smaller elements, the stepwise variation nearly approximates the true smooth variation. Based on several researchers' experience [16, 33, 35], it is accepted that when studying fluidic movement in a microfluidic system, lumped-element models can provide good results if the fluidic element is concentrated into 10 elements per wavelength at the highest operating frequency.

Therefore, after considering the fundamental characteristics and their relationships to multiple energy domains, we conclude that lumped-element models with ODAEs are appropriate for describing and studying dynamic MEFS behavior involving multiple, coupled energy domains.

### 2.1.2 Fundamental Variables

The formulation of state-determined system models requires the selection of a set of fundamental variables. These variables represent the energy interaction within a system, between systems, and with its environment. In addition, these variables provide the basic definition for lumped-element model energy sources, energy storage, energy dissipation, and energy transformation. A uniform classification of variables associated with power and energy is necessary to conveniently model the system's coupled-energy domain behavior. MEFS modeling and simulation methodologies are always based on analogies between these variables in different energy domains. The principle of energy-conservation provides the basic methodology to characterize and define the fundamental variables in mechanical, electrical, fluid, and thermal energy domains [34].

The physical quantities in multiple energy domains can be viewed as types of single-port element variables: *across* and *through*. These two variables are used to describe the power and energy flow variables respectively. An *across* variable denotes a difference in a physical condition across the terminals of an element. A *through* variable denotes a physical quantity transmitted through the terminals of an element. Table 2.1 lists examples of across and through variables for several energy domains. Note that transmission of a through variable does not necessarily imply motion, as with mechanical through variables of force and torque.

- Mechanical translational system
  The dynamics of mechanical systems are governed by the laws of mechanical energy conservation and are described by Newton's laws of motion. There are two mechanisms for energy storage and one energy dissipation within a mechanical system:
  - kinetic energy —associated with moving elements of finite mass.
  - potential energy —stored through elastic deformation of springlike elements
  - dissipated energy —frictional losses of damper elements

  These three elements form the basis to describe the lumped-element modeling of mechanical translation systems [34].

**Table 2.1** Through and Across Variables for Energy Domains

| Energy Domain | Through Variable | Across Variable |
|---|---|---|
| Electrical | Current $i = \frac{q}{t}$ | Voltage $v = \frac{\lambda}{t}$ |
| Mechanical - Translational | Force $F = \frac{p}{t}$ | Velocity $u = \frac{x}{t}$ |
| Mechanical - Rotational | Moment $M = \frac{h}{t}$ | Angular Velocity $\Omega = \frac{\Theta}{t}$ |
| Fluidic | Fluid Flow $Q = \frac{V}{t}$ | Pressure $P = \frac{\Gamma}{t}$ |
| Thermal | Heat Flow $q = \frac{\Psi}{t}$ | Temperature $T$ |

$q$ - electric charge
$p$ - translational momentum
$h$ - angular momentum
$V$ - volume
$\Psi$ - heat energy
$\lambda$ - flux linkage
$x$ - translational displacement
$\Theta$ - angular displacement
$\Gamma$ - pressure momentum

By using the through variables and the across variables in Table 2.1, the increment of energy ($\Delta E$) performed on the mechanical system by the external sources over the period $dt$ can be expressed in the following three forms:

$$\Delta E = F(udt) = Fdx = d\varepsilon_{potential} \quad (2.2)$$
$$\Delta E = u(Fdt) = udp = d\varepsilon_{kinetic} \quad (2.3)$$
$$\Delta E = (Fu)dt = (uF)dt = d\varepsilon_{dissipated} \quad (2.4)$$

These three forms illustrate three basic fundamental elements in lumped-element models of mechanical systems—the spring, mass, and damper elements.

- Mechanical rotational systems
  In rotational systems, power is transmitted and energy is stored by rotary motion about a single axis. By using the across variable and the through variable in Table 2.1, the rotational energy ($\Delta E$) crossing the system boundary over time $dt$ can be expressed in the following three forms:

$$\Delta E = M(\Omega dt) = Md\Theta = d\varepsilon_{potential} \quad (2.5)$$
$$\Delta E = \Omega(Mdt) = \Omega dh = d\varepsilon_{kinetic} \quad (2.6)$$
$$\Delta E = M\Omega dt = \Omega Mdt = d\varepsilon_{dissipated} \quad (2.7)$$

The mechanical rotational power flow may make a system stored energy change in angular displacement $d\Theta$ associated with a rotational spring, or may result in change in angular momentum $dh$ associated with a rotational mass or inertia, or may be dissipated by conversion of rotational work to heat in a rotational damper.

- Electrical systems

  In the electrical energy domain, the across variable and the through variable are a pair of wires: current ($i$) and voltage drop ($v$), as shown in Table 2.1. The electrical energy passing through a system boundary in time $dt$ may also be written in terms of the following three forms:

$$\Delta E = i(vdt) = id\lambda = d\varepsilon_{potential} \tag{2.8}$$

$$\Delta E = v(idt) = vdq = d\varepsilon_{kinetic} \tag{2.9}$$

$$\Delta E = ivdt = vidt = d\varepsilon_{dissipated} \tag{2.10}$$

  Electrical energy crossing a system boundary may result in a change in the magnetic flux $d\lambda$ associated with the system through electromagnetic energy storage in inductors, a change in the total charge $dq$ in a system associated with the electrostatic energy storage in capacitors, or dissipation in resistors through the generation of heat with no electrical energy storage.

- Fluidic systems

  Although complex fluidic phenomena involve flow variables in continuous functions of both space and time, based on the conclusion in Subsection 2.1.1, MEFS can be adequately modeled with lumped-element models. The power flow through a port into a fluid system can be expressed with the through variable and the across variable: volume fluid flow rate $Q$ and fluid pressure drop $P$. As shown in Table 2.1, the volume $V(t)$ represents the total volume of fluid passing through the port over a given time period. The pressure momentum $\Gamma(t)$ is the time integral of pressure, which is analogous to the momentum in mechanical systems. The increment in energy passing through a fluid port in time $dt$ may be written in the following three forms:

$$\Delta E = Q(Pdt) = Qd\Gamma = d\varepsilon_{potential} \tag{2.11}$$

$$\Delta E = P(Qdt) = PdV = d\varepsilon_{kinetic} \tag{2.12}$$

$$\Delta E = PQdt = PQdt = d\varepsilon_{dissipated} \tag{2.13}$$

  Work done by a fluid crossing a system boundary may result in a change in pressure momentum $d\Gamma$ in the system associated with energy storage in a fluid inertance. It also may result in the energy stored in a fluid volume $dV$ associated with fluid capacitance. In addition, it may result in a change on the pressure drop and the flow rate representing the dissipation in a fluid resistance. At this stage, there is no energy storage but fluid energy converts to heat.

- Thermal systems
  Unlike the mechanical, electrical, and fluidic systems, the thermal system's power flow is not commonly described as a product of two variables: the across variable and the through variable. As shown in Table 2.1, thermal systems have historically been characterized in terms of the thermal energy $\Psi$ and the heat flow $q$ and the relationships of these variables to temperature $T$.

### 2.1.3 Relationships between Fundamental Variables

Based on the above multiple energy domain definitions, the relationship between two variables (across and through) can be defined by three general element constitutive relations:

$$\text{across variable} \propto \text{through variable} \tag{2.14}$$

$$\text{across variable} \propto \frac{d}{dt}(\text{through variable}) \tag{2.15}$$

$$\text{across variable} \propto \int (\text{through variable})\, t \tag{2.16}$$

For linear element constitutive relations, the across and through variables are related by constants of proportionality that depend on geometric configurations and material properties.

**Table 2.2** Element Constitutive Relations - Energy Dissipation ($\varepsilon_{dissipated}$)

| Energy Domain | Constitutive Relation | Coefficient |
|---|---|---|
| Electrical | $v = R \cdot i$ | R - resistance |
| Mechanical - Translational | $F = b \cdot u$ | b - damping coefficient |
| Mechanical - Rotational | $M = B \cdot \Omega$ | B - damping coefficient |
| Fluidic | $P = R_f \cdot$ | $R_f$ - fluid resistance (Poiseuille's Law) |
| Thermal | $T = R_t \cdot q$ | $R_t$ - thermal resistance (Fourier's Law) |

Constitutive relations form constraints that must be satisfied at every instance in time.

The three general element constitutive relations form templates for phenomenological laws commonly used in many physical disciplines. For instance, the first relation given in (2.14) denotes energy dissipation conditions in multiple energy domains. Specific examples are given in Table 2.2.

The second and third relations given by (2.15) and (2.16) denote energy storage conditions. These energy storage conditions are analogous to the kinetic energy and potential energy in mechanical systems. Specific examples are given in Tables 2.3 and 2.4. The differential element constitutive relations provide basic time-dependent operations, introducing causality via a rate term describing a nonzero delay between stimulus and response.

**Table 2.3** Element Constitutive Relations - Energy Storage ($\varepsilon_{kinetic}$)

| Energy Domain | Constitutive Relation | Coefficient |
|---|---|---|
| Electrical | $v = L\dfrac{di}{dt}$ | L - inductance (Faraday's Law) |
| Mechanical - Translational | $u = \dfrac{1}{k}\dfrac{dF}{dt}$ | k - translational stiffness (Hooke's Law) |
| Mechanical - Rotational | $\Omega = \dfrac{1}{K}\dfrac{dM}{dt}$ | K - rotational stiffness |
| Fluidic | $P = I\dfrac{dQ}{dt}$ | I - fluid inertance |

The constitutive relations given in Table 2.3 share the property that the supplied kinetic or inertial energy is a function of only the through variable.

$$E = \int K \cdot \text{through variable (through variable)}$$

The constitutive relations given in Table 2.4 (with the exception of thermal) share the property that the supplied potential or capacitive energy is a function of only the across variable, i.e.

$$E = \int K \cdot \text{across variable (across variable)}$$

The constitutive relations of the fundamental elements discussed above make several simplifying approximations, such as the time-invariance of physical properties.

**Table 2.4** Element Constitutive Relations - Energy Storage ($d\varepsilon_{potential}$)

| Energy Domain | Constitutive Relation | Coefficient |
|---|---|---|
| Electrical | $i = C\frac{dv}{dt}$ | $C$ - capacitance |
| Mechanical - Translational | $F = m\frac{du}{dt}$ | $m$ - mass (Newton's Second Law) |
| Mechanical - Rotational | $M = J\frac{d\Omega}{dt}$ | $J$ - mass moment of inertia |
| Fluidic | $Q = C_f\frac{dP}{dt}$ | $C_f$ - fluid capacitance |
| Thermal | $q = C_t\frac{dT}{dt}$ | $C_t$ - thermal capacitance |

Nevertheless, these fundamental across and through variables and their constitutive relations are adequate to describe the MEFS dynamic behavior.

Because of the MEFS coupled-energy effect, system energy is always transferred from one energy domain to another, such as the micropump, mechanical energy is converted to fluid movement energy, and the electrical energy in electrowetting array is converted into droplet fluid sample move energy. Another example is Ampere's Law relating electrical current in the presence of a magnetic field and the induced Lorentz force.

$$\boldsymbol{F} = q\boldsymbol{u}\times\boldsymbol{B} \tag{2.17}$$

The process of energy conversion between different domains is known as *transduction*. Figure 2.2 shows the transduction between different energy domains. In addition, elements that convert the energy are defined as *transducers*. Here, we introduce two types of ideal energy transduction elements. These elements can be used to represent the process of energy transmission.

The basic energy transduction processes can be represented by a two-port element, as shown in Figure 2.3. Energy is transferred from one port to another port. Each port has a through variable ($f_1, f_2$) and an across variable ($v_1, v_2$) defined in its own energy domain.

The two-port transducer illustrated in Figure 2.3 identifies a power flow $P_1$ into port 1 and another power flow $P_2$ into port 2:

$$P_1 = f_1V_1, \qquad P_2 = f_2V_2 \tag{2.18}$$

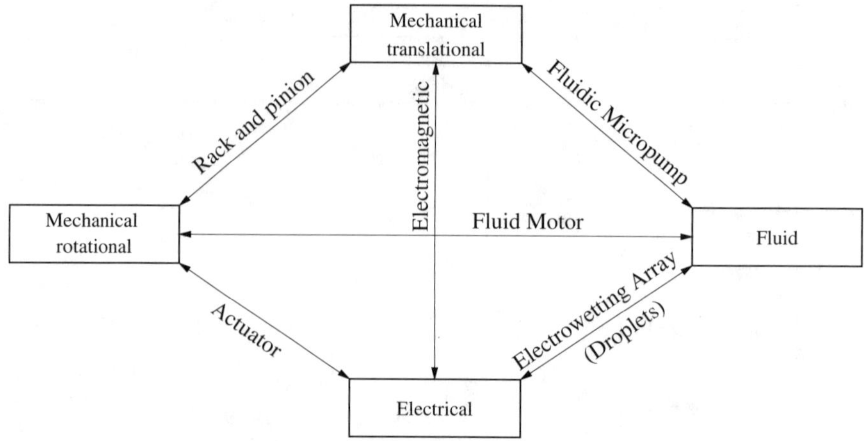


**FIGURE 2.2**
**Transduction between different energy domains**

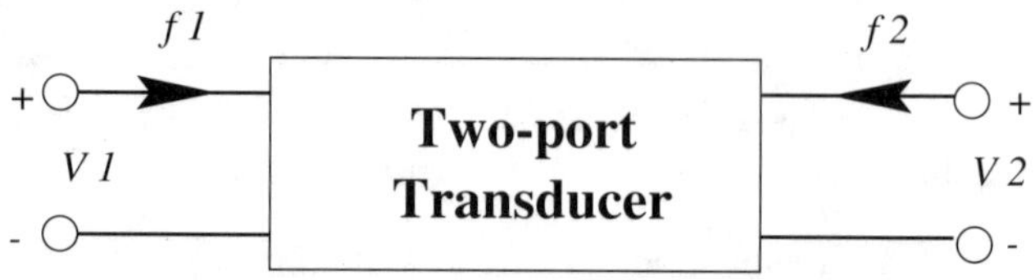


**FIGURE 2.3**
**Two-port transducer consists of two energy domains.**

An ideal energy transduction process is lossless, and power is transmitted with no energy storage or dissipation associated with the transduction process. Therefore, the net instantaneous power sums to zero for all time $t$:

$$P_1(t) + P_2(t) = f_1 V_1 + f_2 V_2 = 0 \tag{2.19}$$

Additionally, the relationship between the across variable and the through variable are represented by constant coefficients and are linear. The most general linear relationship may be written in the following matrix form

$$\begin{bmatrix} V_1 \\ f_1 \end{bmatrix} = \begin{bmatrix} c_{11} & c_{12} \\ c_{21} & c_{22} \end{bmatrix} \begin{bmatrix} V_2 \\ f_2 \end{bmatrix} \tag{2.20}$$

where $c_{11}$, $c_{12}$, $c_{21}$, and $c_{22}$ are constants that depend on the particular transducer. By appropriately selecting these constants, (2.20) yields two possible ideal two-port transducers:

- Transforming Transducer

$$c_{12} = c_{21} = 0 \quad and \quad c_{22} = -\frac{1}{c_{11}}$$

- Gyrating Transducer

$$c_{11} = c_{22} = 0 \quad and \quad c_{21} = -\frac{1}{c_{12}}$$

These two general solutions are the only nontrivial solutions for the ideal two-port transduction. Many physical transduction elements cannot be modeled directly with the ideal two-port element. They may have energy dissipation and storage phenomena associated with the transduction. Therefore, additional lumped-element models must also be incorporated into the transduction system.

### 2.1.4 Kirchhoffian Networks

Once a collection of fundamental elements have been defined, the elements can be connected in a network to model larger physical systems. These systems have more complex and/or distributed dynamical behavior. The interconnection of elements imposes constraints on the variation of system variables. These variables are described via a set of equilibrium equations, also called equations of motion, where motion refers to a change in a physical variable. Equilibrium equations describing the energy interactions between elements can be systematically generated using *continuity conditions* and *compatibility conditions*. Continuity and compatibility conditions are instances of the law of conservation of energy.

Continuity conditions conserve the transmission of certain physical properties with respect to a vertex or node. The sum of such physical properties having a direction of transmission incident to the node must equal the sum of physical properties having a direction of transmission emanate from the node. Table 2.5 lists examples of continuity conditions for several energy domains.

Compatibility conditions conserve the association of certain physical properties with respect to a closed path or loop. The sum of such physical properties around a closed path must be zero. In other words, the accumulated influence of conditions encountered in traversing a closed path of network elements must have no net effect. Compatibility, also called connectedness, conditions reflect various physical constraints of geometry, kinematics, and potential fields. Table 2.6 lists examples of compatibility conditions for several energy domains.

The application of the continuity and compatibility conditions generates a set of equilibrium equations. The equilibrium equations are often formed as a set of second-order ordinary differential and algebraic equations, due to the basic definitions of the

**Table 2.5** Continuity Conditions

| Energy Domain | Continuity Conditions | Conservative Principle |
|---|---|---|
| Electrical | $\sum_{node} i = 0$ | Kirchhoff's Current Law |
| Mechanical - Translational | $\sum_{node} F = 0$ | D'Alembert Law |
| Mechanical - Rotational | $\sum_{node} M = 0$ | D'Alembert Law |
| Fluidic | $\sum_{node} Q = 0$ | Conservation of Mass |
| Thermal | $\sum_{node} q = 0$ | First Law of Thermodynamics |

general element constitutive relations given in (2.14), (2.15), and (2.16). Note that (2.21) is called the *configuration form* of the equilibrium equations; $\boldsymbol{q}$ is a *generalized coordinate* denoting any set of physical quantities that can collectively define system state, or configuration.

$$\ddot{\boldsymbol{q}} = \boldsymbol{f}(\boldsymbol{q}^T, \dot{\boldsymbol{q}}^T, t) \tag{2.21}$$

$$\begin{aligned} \ddot{q}_1 &= f_1(q_1, q_2, \cdots, q_n, \dot{q}_1, \dot{q}_2, \cdots, \dot{q}_n, t) \\ \ddot{q}_2 &= f_2(q_1, q_2, \cdots, q_n, \dot{q}_1, \dot{q}_2, \cdots, \dot{q}_n, t) \\ &\vdots \\ \ddot{q}_n &= f_n(q_1, q_2, \cdots, q_n, \dot{q}_1, \dot{q}_2, \cdots, \dot{q}_n, t) \end{aligned}$$

The second-order differential and algebraic equations can be combined into a smaller set of equations of higher order

$$y^{(n)} + a_1 y^{(n-1)} + \cdots + a_{n-1}\dot{y} + a_n y = b_0 u^m + b_1 u^{m-1} + \cdots + b_{m-1}\dot{u} + b_m u \quad m \le n$$

or expanded into a larger set of equations of lower (first) order.

$$\dot{\boldsymbol{x}} = \boldsymbol{f}(\boldsymbol{x}, \boldsymbol{u}) \tag{2.22}$$

$y$ is the system output, $u$ is the system input, $a_i$ is the state coefficient, and $b_i$ is the input coefficient. $\boldsymbol{x}$ is a state vector $\boldsymbol{x} = [x_1, x_2, \ldots, x_n]^T$ and $\boldsymbol{u}$ is a input vector

**Table 2.6** Compatibility Conditions

| Energy Domain | Compatibility Conditions | Conservative Principle |
|---|---|---|
| Electrical | $\sum_{closed\ path} v = 0$ | Kirchhoff's Voltage Law |
| Mechanical - Translational | $\sum_{closed\ path} x = 0$ $\sum_{closed\ path} u = 0$ | Geometric Kinematics |
| Mechanical - Rotational | $\sum_{closed\ path} \theta = 0$ $\sum_{closed\ path} \Omega = 0$ | Geometric Kinematics |
| Fluidic | $\sum_{closed\ path} P = 0$ | $\nabla \times \boldsymbol{F_P} = 0$ |
| Thermal | $\sum_{closed\ path} T = 0$ | First Law of Thermodynamics |

$\boldsymbol{u} = [u_1, u_2, \ldots, u_n]^T$. If $u$ is defined as an $n$-dimensional state-space, $\boldsymbol{f}$ defines a vector field on $u$ with $\boldsymbol{x}(t)$ defining a specific trajectory or orbit in the plane having tangent $\boldsymbol{f}(\boldsymbol{x})$ at $\boldsymbol{x}$. If the mathematical state vector $\boldsymbol{x}$ possesses the property of invariance, it also denotes a physical vector.

(2.22) can be reformulated using the Taylor series expansion and neglecting higher-order terms yields the linear approximation

$$\dot{\boldsymbol{x}} = \left.\frac{\partial \boldsymbol{f}(\boldsymbol{x},\boldsymbol{u})}{\partial x_1}\right|_{\substack{\boldsymbol{x}=\boldsymbol{x}_0\\ \boldsymbol{u}=\boldsymbol{u}_0}} x_1 + \left.\frac{\partial \boldsymbol{f}(\boldsymbol{x},\boldsymbol{u})}{\partial x_2}\right|_{\substack{\boldsymbol{x}=\boldsymbol{x}_0\\ \boldsymbol{u}=\boldsymbol{u}_0}} x_2 + \cdots + \left.\frac{\partial \boldsymbol{f}(\boldsymbol{x},\boldsymbol{u})}{\partial x_n}\right|_{\substack{\boldsymbol{x}=\boldsymbol{x}_0\\ \boldsymbol{u}=\boldsymbol{u}_0}} x_n$$
$$+ \left.\frac{\partial \boldsymbol{f}(\boldsymbol{x},\boldsymbol{u})}{\partial u_1}\right|_{\substack{\boldsymbol{x}=\boldsymbol{x}_0\\ \boldsymbol{u}=\boldsymbol{u}_0}} u_1 + \left.\frac{\partial \boldsymbol{f}(\boldsymbol{x},\boldsymbol{u})}{\partial u_2}\right|_{\substack{\boldsymbol{x}=\boldsymbol{x}_0\\ \boldsymbol{u}=\boldsymbol{u}_0}} u_2 + \cdots + \left.\frac{\partial \boldsymbol{f}(\boldsymbol{x},\boldsymbol{u})}{\partial u_n}\right|_{\substack{\boldsymbol{x}=\boldsymbol{x}_0\\ \boldsymbol{u}=\boldsymbol{u}_0}} u_n$$

rewritten as

$$\dot{\boldsymbol{x}} = \boldsymbol{A}\boldsymbol{x} + \boldsymbol{B}\boldsymbol{u} \quad \text{state equation} \tag{2.23}$$

$\boldsymbol{A}$ is the state matrix and $\boldsymbol{B}$ is the input matrix.

Additionally, with the linearity approximation, any system variable can be represented as a combination of state variables and system inputs, yielding

$$\boldsymbol{y} = \boldsymbol{C}\boldsymbol{x} + \boldsymbol{D}\boldsymbol{u} \quad \text{output equation} \tag{2.24}$$

$$
\begin{aligned}
\dot{y}_1 &= c_{11}x_1 + \cdots + c_{1n}x_n + d_{11}u_1 + \cdots + d_{1m}u_m \\
\dot{y}_2 &= c_{21}x_1 + \cdots + c_{2n}x_n + d_{21}u_1 + \cdots + d_{2m}u_m \\
&\vdots \\
\dot{y}_p &= c_{p1}x_1 + \cdots + c_{pn}x_n + d_{p1}u_1 + \cdots + d_{pm}u_m
\end{aligned}
$$

$\boldsymbol{y}$ is the output vector, $\boldsymbol{C}$ is the output matrix, and $\boldsymbol{D}$ is the direct transmission matrix. The system dynamical modeling using bond graphs typically generates state-space equilibrium equations.

Kirchhoffian network theory is not the only approach to formulating a set of equilibrium equations. For instance, Hamilton's principle using energy and coenergy form the basis for an alternate formulation, called Lagrangian equations. These network theories formulate the dynamic modeling of MEFS by using the fundamental variables of different energy domains. By solving these equations analytically or numerically, the system behavior can be studied.

### 2.1.5 The Equivalent Circuit Modeling Method

Because of the coupling effect between different energy domains, MEFS models must be simulated simultaneously. It is very hard to avoid convergence problems arising from coupling of heterogeneous simulators. It is beneficial to model all MEFS energy domain phenomena with a common description language, and simulate the complete system dynamic behavior by one simulator. The electrical analogy method has been used in acoustics for transducer modeling for a long time [36]; this method consists of describing a system by a network of mechanical impedances, possibly nonlinear, and subsequently carrying out an analysis with electrical simulation tools such as $SPICE$ [37] and $SABER$ [38].

The process of building an equivalent network consists of subdividing the complete device structure into lumped elements, and each element is then described on the basis of analogies between relevant physical parameters of the phenomenon and electrical parameters. Table 2.7 shows the electrical analogs for fluidic and mechanical parameters.

The equivalent circuit approach can be applied to many coupled-energy domain problems, and its use is strongly supported by modern electric network theory. In addition, the equivalent circuit approach is particularly useful for the analysis of systems consisting of complex structural elements and coupled subsystems with multiple ports [39]. Moreover, the equivalent circuit approach appears to be a better visualization of the system, and it is very useful in the field of microsystems owing to the strong coupling between all constitutive parts of the system [40]. Section 5.1.1.2 shows its application on a microelectromechanic device: a comb-drive microresonator.

**Table 2.7** Fluidic and Mechanical Parameters and Electrical Similarity

| Mechanical parameters | | Fluidic parameters | | Electrical parameters | |
|---|---|---|---|---|---|
| $u$ | velocity | $\Phi$ | flow rate | I | current |
| F | force | P | pressure | v | voltage |
| $p$ | translational momentum | V | volume | Q | charge |
| $m$ | mass | m/$S^2$ | | L | inductance |
| $\alpha$ | friction coefficient | $\alpha/S^2$ | | R | resistance |
| $k$ | stiffness | k/$S^2$ | | C | capacitance |

Note: $S$ is the area normal to the direction of motion.

However, the choice of limited circuit elements in the $SPICE$ library reduces the flexibility of the equivalent circuit approach. Additionally, it is difficult to find the equivalent circuit when the components become complex, especially when higher-order differential equations are needed to model nonlinear behavior [41].

The primary difficulties with the equivalent circuit approach may be avoided by the use of modern hardware description languages. For example, VHDL-AMS can be used to provide a form to study the dynamic behavior of the system. This form is associated with ODAEs. The detail VHDL-AMS modeling and simulation approach will be discussed in Section 3.1.1.

## 2.2 MEFS System-level Modeling and Simulation

MEFS system-level modeling includes the architectural-level stochastic modeling and biomedical/chemical process flow modeling. Stochastic modeling and simulation provide a level of abstraction for studying architectural performance issues, and statistical methods are used to analyze the results. Queuing theory is used to model resource contention where service requests exceed service capacity and processing must be interrupted and delayed [42]. Process flow modeling and simulation provide an additional level of abstraction for studying biomedical application execution issues. In addition, performance simulation models use a functional modeling style, commonly called macro modeling [41]. Macro modeling emphasizes computational execution efficiency by describing the "black box" behavior of a component. The input/output transformation is described, but information about how the transformation is performed is not described. The macro models focus on performance related information, such as the throughput, capacity, and overall execution times.

Several unique system-level modeling factors need to be studied. These factors are necessary to describe the system-level bio/chemical application behavior and MEFS architecture. They are also critical parts of a MEFS modeling and simulation language. Factors that relate to the MEFS system-level modeling are defined in Section 2.2.1, which consists of multiple system description perspectives, object-oriented and dynamic data structures, and the system specification capacity. In Section 2.2.2, factors reflecting the system-level simulation are presented. They include the time-advanced mechanism [43] which is useful for keeping track of the simulated time, and system design scalability. Moreover, due to the runtime nondeterminism of stochastic systems, MEFS system-level statistical analysis factors are defined in Section 2.2.3.

## 2.2.1 MEFS System-level Modeling

### 2.2.1.1 MEFS Behavior Modeling Perspectives

System behavior can be modeled with either discrete or continuous representation. These representations are commonly classified into three categories, named "worldviews," which are commonly used by modeling and simulation languages to conceptualize a domain or system [44]:

- Discrete Event-Scheduling
- Discrete Process-Interaction
- Continuous

Though the event-scheduling worldview is convenient for describing transformations for low-level digital systems, the process-interaction worldview is often convenient for describing the queuing nature of higher-level stochastic systems. Additionally, the process-interaction worldview can be used more easily to describe object-oriented system behavior [45]. Multiple worldviews can be combined, allowing portions of the dynamical behavior to be described by a discrete modeling paradigm and other portions of the dynamical behavior to be described by a continuous modeling paradigm.

Although the basic MEMS dynamic behavior is described using event-scheduling, continuous or both perspectives, this modeling paradigm can not directly represent MEFS behavior. MEFS behavior not only requires a combination of discrete and continuous perspectives, but it also requires the discrete representation including event-scheduling, process-interaction, and a combination of both. For instance, the event-scheduling perspective is necessary to model a microfluidic sample arrival event. The continuous perspective is used to describe a thermal reaction involving solution mixtures, and the energy-conservative queuing nature of a biochemical DNA analysis system must be described with a process-interaction perspective [46].

#### 2.2.1.2 Object-oriented and Dynamic Data Structure

In contrast to the MEMS processor-oriented modeling perspective, the MEFS system level design mainly focuses on the change of fluidic sample characteristics. Therefore, the process-interaction worldview is popularly adopted to describe this fluidic-sample-oriented MEFS behavior. In addition, in order to more effectively represent the features of each fluidic sample in MEFS, a complex but flexible data structure is necessary. The features are defined in the following.

- Fluidic Sample Property
  This item presents fluidic sample physical and chemical features, such as the fluidic sample volume and the sample temperature, etc. These features may be changed during the fluidic sample processing period.

- System Resource Utilization
  This item records the status of each fluidic sample using system resources. For instance which processor is used by that fluidic sample? Which channel is used to deliver that fluidic sample from the storage buffer to the processor or from the processor to outlet?

- Fluidic Sample Simulation Clock
  This item records the simulated time value of each process event for a certain fluidic sample. For example, the time value when that fluidic sample arrives at the input to the handling system, the time value when that fluidic sample arrives at a storage buffer, the thermal reaction time for that fluidic sample, etc.

By definition, stochastic systems exhibit runtime nondeterminism; any particular simulation is simply one observation of the random behavior. Thus, data structures associated with random variables cannot generally be predefined at model development or instantiation, and dynamic data structures are required that can be modified during runtime of the model, i.e. simulation. A dynamic data structure, such as a linked-list [43], can be an effective data representation for a set of objects or data values, where set membership can vary during simulation by creating and destroying objects. For instance, a linked-list is often used to represent a set of fluidic samples waiting for service or a set of jobs waiting for execution [47]. In addition, this dynamic data structure can describe more complicated queuing behavior, involving priorities, preemptions, redistributions, and terminations [27].

#### 2.2.1.3 User-definable Behavior

One of the most effective ways to address the difficulties of MEFS design complexity is to create abstractions at the system level. These abstraction highlight relevant system characteristics and deemphasize or hide all other information. They also reveal how a designer views the intent and operation of a complex system. Thus system performance modeling languages are required to provide a basic set of predefined functions and behaviors to construct application-specific, user-definable abstractions [48]. In addition, because of the complexity of MEFS architecture, the different functional blocks can be connected to each other sequentially or in parallel, as shown in Figure 2.4, either the sequential process contains the concurrent procedure, or the concurrent procedure embodies the sequential process.

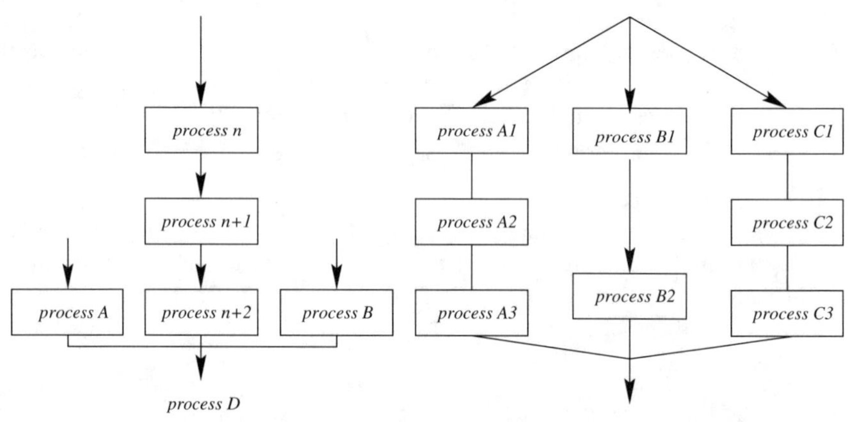


**FIGURE 2.4**
**Mixed sequential and concurrent execution includes either the sequential process containing the concurrent procedure, or the concurrent procedure embodying the sequential process.**

This aspect of system-level modeling can be realized using a variety of constructs and statements supporting both sequential and concurrent executional semantics for procedural and parallel tasking and methodologies, respectively. In this manner, a system performance modeling language is molded to fit an application rather than the counter situation of contorting an application to fit an inflexible system performance modeling language.

#### 2.2.1.4 System Specification

Based on the state-of-the-art top-down design methodology, MEFS modeling and simulation present new requirements for the initial conceptual design at the system-level and supporting computer-aided tools. This system-level specification is always represented with a sequential process. At the beginning of a design flow, the accurate executable system design specification at system level is created. It helps validate the MEFS bio/chemical application and functionality, and creates the system application performance model. With the top-down decomposition, the application-level design is slowly refined into different functional blocks depending on the system structure. At the end, the detail functional units are implemented which match the MEFS component capability design, and fulfill the system application requirement.

### 2.2.2 MEFS System-level Simulation

#### 2.2.2.1 Time-Advanced Mechanisms

Due to the stochastic nature of MEFS application behavior, a variable is necessary to keep track of the current value of simulated time when the simulation proceeds. This variable is called the *simulation clock*. It is also useful to advance simulated time from one value to another for the event-scheduling worldview. In addition, it is necessary to synchronize events in the simulation. Clocks order events in time so that parallel events are properly modeled by simulator on a sequential computer. There is generally no relationship between simulated time and the time needed to run a simulation on a computer.

#### 2.2.2.2 Scalability of Simulation

Because of the hierarchical structure of MEFS, MEFS simulation requires the study of design scalability. Existing MEMS hierarchical modeling and simulation techniques focus on low-level components. The entire MEMS component as a single behavior entity forms the top level of hierarchy, and the constituent MEMS elements, such as plate masses and beam springs, form the hierarchical lower level [25]. However, this approach is not efficient for MEFS. A scalable methodology for MEFS must handle heterogeneous, multiple-component systems, and address complex fluidic-application and mixed-level component simulation. It is important to investigate how the performance of a microliquid handling system architecture scales with increasingly complex chemical and biological analyses, and what types of biomedical applications can be practically miniaturized via microfluidic molecular processing. In addition, it is also necessary to investigate how the performance of the microliquid handling system scales with advances in constituent microfluidic device technology. Therefore, the MEFS system-level modeling and simulation languages must possess a hierarchical scalable-design capacity.

### 2.2.3 Statistical Analysis Capacity

The purpose of simulation is to imitate the operation of a real-world system, and then to use the resulting simulation output data to infer the real-world system functionality and performance. MEFS high-level system performance models are generally stochastic because either the system is too complex to be analytically characterized, design details are unknown, or overall performance depends on ambient factors that are nondeterministic [49]. Stochastic systems dynamically vary over time because the system operation is dependent on one or more random variables. Hence, the resulting simulation output data exhibit random variability. Consequently, the statistical analysis approach is very important [27]. Statistical analyses require the language capacity to compile various usage information during system execution to estimate the mean, variation, correlations, and confidence intervals of the sampled random results. Probabilistic and statistical analysis also require multiple data types, powerful mathematical resources (function libraries), and operating system storage (file) input/output.

## 2.3 Conclusion

The lumped-element models with ODAEs are appropriate to describe the MEFS circuit-level dynamic behavior coupled with multiple energy domains. An equivalent circuit approach can be used for MEFS circuit level device modeling and simulation, and its primary difficulties may be avoided by the use of modern hardware description languages. In addition, the MEFS system-level hierarchical modeling and performance evaluation require the description capacity of simulation languages for MEFS system-level hierarchical modeling, simulation, and statistical analyses.

System design includes two phases: the *conceptual phase* and the *product-level phase* [50]. Based on these two phases, there are two different types of behavior description capacity requirements. In the *conceptual phase* of a new device, the design objective is to build the practical configuration of the system. Accurate but abstract information is necessary. After system analysis and design refinement, detailed functional requirements for each device are obtained. System design turns into the second phase, *product-level phase*. In this phase, detailed physical behavior and parasitic phenomena are studied. Right now, our MEFS system-level research mainly focuses on the behavior modeling at the conceptual phase. Therefore, it is assumed that the MEFS system-level performance is the accumulation of performance of each microfluidic component. For instance, the overall flow rate $\Phi$ of two parallel connected micropumps is the sum of each micropump flow rate $\Phi_1, \Phi_2$: $\Phi = \Phi_1 + \Phi_2$. In particular, it is assumed that there are no parasitic phenomena existing between microfluidic components. For example, because the micropumps are independent even though they are interconnected, their behavior can be studied independently.

# *Chapter 3*

## *SystemC-based Hierarchical Design Environment*

Simulation applications are becoming major software systems, at par with the complexity of databases or operating systems. This can be attributed to the increasing complexity of the systems being analyzed and the need to conduct the analysis across a wide range of abstraction levels. System design is evolving into a multidisciplinary field requiring expertise in electrical, mechanical, and chemical engineering, as well as in computer science and manufacturing technology. This broad scope makes it nearly impossible to understand complicated factors influencing and limiting system performance. It often requires the collaborative efforts of several teams and organizations, typically using different modeling languages and simulators.

Traditionally, for complex heterogeneous system design, several modeling languages and simulators are used to support various phases of system specification, architectural design, and functional unit design. Performance modeling languages such as SIMSCRIPT II.5, SLAM II, and general purpose software programming languages such as C and Ada are used for the high-level stochastic architectural design and the biomedical/chemical process flow simulation [27], [51]. On the other hand, logic modeling languages, for instance VHDL/VHDL-AMS and Verilog, are used for low-level functional unit design [52]. In addition, array modeling languages such as Matlab are used to model the dynamic behavior of the system. One of the major problems in simulation is the need for human intervention during the information transfer between different simulation languages and simulators. For instance, the prevalent top-down system design methodology starts with C or C++ to model the system at the system level. This step is used to verify basic design concepts and algorithms. Following this step, the hardware components of the system are manually converted from a C/C++ model to VHDL or Verilog descriptions for the actual hardware implementation. The discontinuous design flow requires multiple levels of repetitive verification. In addition, different tools and host platforms can lead to problems in misinterpretation of concept specifications, misunderstanding of data, and loss of information during translation. Such interoperability problems increase the probability of errors, redesign costs, and design cycle times.

Thus, it is necessary to construct a hierarchical system modeling and simulation environment using a common system description language and associated simulation engine, rather than multiple languages and simulators that span the different abstraction level. In this chapter, a hierarchical modeling and simulation environment based on SystemC is presented. In Section 3.1, we examine the suitability of several simulation languages for MEFS hierarchical design. These languages include VHDL/VHDL-AMS, SLAM, C/C++, Matlab, and SystemC. Next, SystemC is proposed as a potential candidate for complete system modeling and simulation. In Section 3.2, a hierarchical modeling and simulation environment based on SystemC is presented. The architecture of the environment and the associated functional packages are discussed. Finally, conclusions are presented in Section 3.3.

## 3.1 Suitability of Modeling Languages for Hierarchical Design

### 3.1.1 VHDL-AMS Suitability for Circuit-level Modeling and Simulation

The VHSIC (Very High Speed Integrated Circuits) Hardware Description Language VHDL was originally designed to describe the structure and behavior of discrete time systems [43]. Recently, VHDL has been extended to enable descriptions of continuous-time systems. The combination of discrete- and continuous-time language constructs are collectively referred to as VHDL-AMS, where the suffix AMS stands for analog and mixed signal. VHDL-AMS provides, for the first time, an integrated capability via a single simulation language for describing discrete systems, continuous systems, and their combinations [53].

VHDL-AMS provides additional semantics to describe continuous-time systems with behaviors governed by a set of simultaneous equations. These sets of equations include simultaneous ordinary differential and algebraic equations (ODAEs). ODAEs are described in a denotational style using "simultaneous statements". These statements define conditions or relations that must always hold over time. Whereas VHDL uses an imperative dynamical model of loosely-coupled concurrent processes communicating by signals, VHDL-AMS uses a declarative dynamical model of tightly-coupled simultaneous relations influencing each other's solution by linked unknowns.

Unknowns are continuous, analytical functions of time, determined by repeatedly invoking an "analog solver" to solve the equations over a series of intervals denoting a period of time. The unknowns are called *quantities* and constitute a new class of objects in VHDL. Quantities are like signals and they denote a waveform or a time series of values. However, quantities are very different from variables or signals in that they do not participate in assignment statements; quantities take their values as a result of solving the set of simultaneous ODAEs. This update mechanism is unique to quantities and is the principal reason for their role as a distinct object class.

Various operations associated with a quantity that yields related quantities, such as its derivative and integral, are defined in VHDL-AMS as predefined attributes. For example, the time derivative of a quantity representing translational velocity, named `VELOCITY`, is denoted by `VELOCITY'DOT`, where `DOT` is a predefined attribute for differentiation. The time derivative of the quantity `VELOCITY` is automatically computed based on the sequence of values of `VELOCITY` computed by the analog solver at various analog solution points (ASPs).

The ODAEs governing MEFS possess a global structure reflecting the fact that physical systems obey laws of conservation of energy. A set of equations exhibiting such a global structure is called a *conservative* set of ODAEs. To facilitate generating conservative ODAEs, VHDL-AMS provides a set of language constructs, involving *terminals*, *natures*, and *branch quantities*.

Constitutive relations are defined using branch quantities and simultaneous statements. Branch quantities possess additional semantics to reflect the additional constraints that govern conservative simultaneous ODAEs. There are two kinds of branch quantities: *across* and *through*. Examples of across and through variables are given in Table 2.1.

As defined on Table 2.5 in Chapter 2, an across physical concept in VHDL-AMS obeys a conservation of energy law in which the summation of across quantities around any closed path (loop) equals zero.

$$\sum_{\text{CLOSED PATH}} \text{ACROSS QUANTITIES} = 0$$

Across quantities are often gradient potentials within a given energy domain and thus they can be viewed as energy.

As defined on Table 2.6 in Chapter 2, a through physical concept in VHDL-AMS obeys a conservation of energy law in which the summation of through quantities at any node (point) equals zero.

$$\sum_{\text{NODE}} \text{THROUGH QUANTITIES} = 0$$

Through quantities are analogous to the movement of state under an applied gradient potential field within a given energy domain; thus they can be viewed as power flow.

Note that the product of across and through quantities can, but need not, denote power. The through quantity for thermal energy domain (heat flow) has units of power. Across and through quantities are computationally defined to ensure correct formulation of the system of dynamical equations.

The types of branch quantities are indirectly defined via the definition of the *terminals*. Terminals provide connection points for conservative equations. Terminals hold no values; their only role is to facilitate the formation of conservative equation sets that, in turn, constrain the associated branch quantities. `ELECTRICAL` and `TRANSLATIONAL` are not types; they are *natures*. Natures can be considered similar to types—both of them define data templates or structures for objects. The nature declarations for `ELECTRICAL` and `TRANSLATIONAL` are supplied by the packages `ELECTRICAL_SYSTEMS` and `MECHANICAL_SYSTEMS`, respectively. Natures can be viewed as abstract descriptions of an energy domain in that they define the fundamental energy/power flow concepts of a physical discipline.

#### 3.1.1.1 Common Declarations

To exchange MEFS component models as intellectual property, an infrastructure is required establishing common terminology, declarations, styles, and practices [53]. The infrastructure supports the common use of specialized applications of a base linguistics. This increased level of specificity is required to ensure MEFS models can be interoperated. Candidate elements of a common VHDL-AMS modeling infrastructure for MEFS are listed below:

- Package Architecture and Format,
- Types and Subtypes,
- Natures and Subnatures,
- Physical Constants,
- Simulation Controls, and
- Design Unit Formats.

Package architecture refers to how the physical concepts and principles of the domain of MEFS can be mapped to a set of VHDL-AMS packages. Packages provide the ability to group declarations together into an encapsulation to represent a convenient abstraction. Additionally, packages can use the abstraction for easily creating new component models. The base package `ENERGY_SYSTEMS` contains declarations common across energy domains and is used by the derived packages representing individual energy domains and disciplines. The VHDL-AMS package architecture provides the advantage of a simple strategy for constructing a component model of a MEFS involving multiple energy domains; the appropriate packages representing each energy domain are included via use clauses. The proposed VHDL-AMS package architecture also provides separate tools for each major engineering discipline to develop the data representations and operations respectively. Such declarations can reflect established terminology and notational conventions. In addition, these declarations are consistent with recognized design practices and analysis methods pertinent to each discipline.

In considering a common typing/subtyping structure for modeling MEFS, it is important to note that VHDL-AMS quantities are restricted to be of a floating-point type so they can represent analytical functions of time. This restriction implies that across types and through types defined via nature declarations must also be floating point types.

The new types are declared as subtypes of the common parent or base floating point type REAL. In addition, these types facilitate the implementation of numerical algorithms for solving ODAEs. Moreover, these types can avoid the complexity of providing a large number of overloaded operators to support dimensional analysis. The new types for across and through branch quantities can be used to declare a common set of natures. It is important to point out that a given energy domain may not necessarily be characterized by a single nature declaration, since the definitions of across and through physical phenomena may be satisfied in multiple ways. Compatibility relations defined in Table 2.5 for mechanical systems offer a simple example. In addition to displacement differentials, the time derivative of displacement also has physical significance in mechanics, namely velocity.

### 3.1.1.2 Micropump lumped-element nodal modeling

In order to illustrate the VHDL-AMS modeling and simulation methodology, lumped-element constitutive relations and conservative Kirchhoffian network theory are used to build a micropump model [18]. The resulting set of simultaneous ODAEs are described using VHDL-AMS.

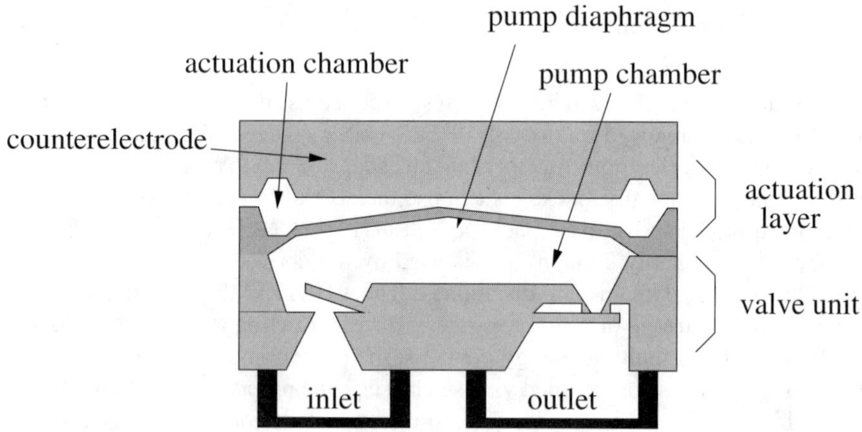


**FIGURE 3.1**
**Schematic view of an electrostatically-driven diaphragm pump [4].**

Micropumps provide the pressure gradient for moving liquid in channels, reservoirs, and chambers. Micropumps typically consist of an actuation unit. Piezoelectric, thermopneumatic and electrostatic actuation are often used [4]. The check microvalves control fluid movement through the inlet and outlet ports. Microchannels connect the inlet and outlet ports to the large channel network.

Figure 3.1 shows the schematic view of an electrostatically-driven diaphragm bi-directional micropump [4]. The operation of the micropump can be divided into two stages. During the pump stage, the condenser is biased, the pump diaphragm deflects upward, and an amount of fluid ($\Delta V$) flows in the inlet port to the pump chamber, the input microvalve is open and the output microvalve is closed. During the supply stage, the biasing voltage is turned off, the pump diaphragm is released to return to its equilibrium position, the $\Delta V$ fluid is ejected out of the pump chamber to the outlet port, the output microvalve is opened and the input microvalve is closed. Varying the actuation frequency changes the pump direction. The mechanical resonance of the check microvalves causes a phase shift between the movement of the microvalve and the pressure difference driving the fluid.

Three coupled differential equations, (3.1)-(3.3), are used to describe the transient behavior of the electrostatically-driven micropump and the passive check valve movement [16].

$$\dot{p} = \frac{\Phi_{iv}(p, x_{iv}) - \Phi_{ov}(p, x_{ov})}{\frac{dV_o(p)}{dp} - \frac{dV_{gas}(p)}{dp}} \tag{3.1}$$

where

| | |
|---|---|
| $iv$ – inlet valve | $x_{iv}$ – inlet valve displacement |
| $ov$ – outlet valve | $x_{ov}$ – outlet valve displacement |
| $\Phi_{iv}$ – inlet flow | $V_0$ – volume of the pump chamber |
| $\Phi_{ov}$ – outlet flow | $V_{gas}$ – volume of gas bubble |
| $p$ – pressure | |

The dynamics of the microvalves are described by second-order ordinary differential equations. The driving force ($F$) is the product of the pressure difference and the area of the microvalve.

$$m_{iv}\ddot{x}_{iv} + d_{iv}\dot{x}_{iv} + k_{iv}x_{iv} = F_{iv}(t) = S_{iv}(p_1 - p) \tag{3.2}$$

$$m_{ov}\ddot{x}_{ov} + d_{ov}\dot{x}_{ov} + k_{ov}x_{ov} = F_{ov}(t) = S_{ov}(p - p_2) \tag{3.3}$$

where

| | |
|---|---|
| $m$ – effective mass | $d$ – damping constant |
| $k$ – elasticity | $p$ – pressure in chamber |
| $F$ – force | $p_1$ – pressure in inlet |
| $S$ – area | $p_2$ – pressure in outlet |

The micropump is partitioned into five parts: an inlet channel, an inlet valve, a pressure chamber, an outlet valve, and an outlet channel. The five parts are modeled and simulated using VHDL-AMS. Figure 3.2 shows the general VHDL-AMS model for a micropump [54].

| | |
|---|---|
| <pre>library ieee;<br>use<br> work.electrical_system.all;<br><br>use<br> work.fluidic_system.all;<br><br>use<br> ieee.math_real.all;<br><br>entity MICROPUMP is<br>   ....<br><br>end entity MICROPUMP;</pre> | <pre>architecture CONFIGURATION of<br>             MICROPUMP is<br>   ....<br>begin<br> -- VALVE<br> InLetValve  :<br>   entity work.Valve<br>             port map ( .... );<br> OutLetValve :<br>   ...<br> -- InLet Channel<br> InChannelResistor  :<br>   entity work.FluidResistor<br>             port map ( .... );<br> InChannelInductance:<br>   ...<br> -- OutLet Channel<br> OutChannelResistor:<br>   entity work.FluidResistor<br>             port map ( .... );<br> OutChannelInductance:<br>   ...<br> -- Chamber<br> Diaphragm:<br>   entity work.FluidCapacitance<br>             port map ( .... );<br>end architecture CONFIGURATION;</pre> |

**FIGURE 3.2**
**The VHDL-AMS model of a micropump includes five functional parts.**

In summary, VHDL-AMS supports circuit-level modeling and simulation of continuous and discrete systems with conservative and non-conservative semantics. The equations describing the conservative aspects of a system does not need to be explicitly notated by the user. The VHDL-AMS solver automatically verifies the conservation of energy. A micropump is used as a special case study to illustrate the MEFS component modeling with VHDL-AMS.

However, the processor-oriented modeling perspective of VHDL-AMS limits its applicability for MEFS fluidic-sample-oriented analysis. For example, it can not provide a complex and flexible data structure to describe the fluidic sample characteristics. In addition, VHDL-AMS, which supports the conservative ODAEs, may not be suitable to directly describe the more complex fluidic PDAEs modeling requirement for MEFS.

### 3.1.2 VHDL Suitability for System-level Modeling and Simulation[1]

In the following sections, the suitability of VHDL is discussed for supporting an integrated approach for modeling and simulation of system-level discrete, continuous, and combined discrete/continuous dynamical behavior of stochastic systems. At first the description capability of VHDL is discussed for MEFS performance modeling. The next section explains the role of VHDL in stochastic discrete system performance evaluation and presents experimental results. Then, the use of VHDL for continuous system performance evaluation is presented along with supporting experimental results. Finally, an example is given combining stochastic discrete and continuous system performance simulation to illustrate overall VHDL system-level modeling and analysis capacities. A summary of key observations makes up the conclusion.

#### 3.1.2.1 System Performance Modeling Simulation

VHDL supports a broad range of constructs describing detailed logic to abstract algorithm and thus, provides the semantics for supporting an integrated modeling and simulation capability. Work has been reported on investigating applications of VHDL to high-level design, including performance modeling and simulation [55, 56]. These efforts focused on specific representational paradigms, such as Petri nets, and emphasized single energy domain (electronic) systems [57, 58]. Our work focuses on a more general resource-contention representational paradigm and emphasizes coupled-energy domains.

- Worldviews
  The basic runtime or dynamic semantics of the VHDL simulation microkernel uses event-scheduling to record the results of invoking processes and promulgating the results to affect causal behavior. Signals are the primary medium for information exchange between VHDL processes. Though signals provide basic time sequencing for values, signals do not directly possess queuing properties for values [59]. Thus, to model and simulate the process-interaction worldview with VHDL, queuing facilities need to be developed using additional data abstractions and operations. In addition, to model and simulate

[1]Reprinted from Microelectronics Journal, Vol. 31, T. Zhang, A. Dewey, and R. B. Fair, A Hierarchical Approach to Stochastic Discrete and Continuous Performance Simulation Using Composable Software Components, pp. 95-104, Jan. 2000. © 2000, with permission from Elsevier Science.

continuous worldview with VHDL, differential equations with respect to time must be discretized and transformed into corresponding difference equations. In this manner, evaluation of a continuous differential problem is approximated by simulation of a corresponding finite-difference problem.

- User-Definable Behavior
VHDL possesses both concurrent execution semantics to reflect the low-level parallelism of hardware and sequential execution semantics to reflect the high-level ordering of procedures and functions. This description capacity supports exploration of alternative system designs and incremental refinement of a particular system design. Additionally, the basic set of VHDL predefined functions and behaviors can be enhanced by user-defined data types and overloaded operators. User-defined attributes can also be defined for system annotation.

- Dynamic Data Structures
This aspect of system performance modeling can be realized using the pointer data type, called *access type* [59]. The use of access types in VHDL has limitations in that signals may not be declared to be of an access type—signals are statically defined. Only variables can be declared as an access type, which limits their application in global information exchange. Variables are constrained to have scope within a process; shared variables are constrained to have scope across processes within an architecture [60]. Thus, there is a need to extend these base simulation microkernel capacities by additional data abstractions and operations to provide the desired capabilities of dynamic data structures for performance simulation.

- Output Analysis and Report
This aspect of system performance modeling can be realized in VHDL using real data types and arithmetic operators augmented by the VHDL Math package (IEEE Standard 1076.2) for mathematical computations. Files provide a logical interface to the physical storage capabilities of an underlying operating system to support output reporting [60]. Also, shared variables provide a medium for implementing global accounting practices to collect usage data for statistical analysis.

#### 3.1.2.2 Stochastic Discrete-Event Performance Simulation

The salient aspects of stochastic discrete-event performance simulation will be discussed by illustrating the modeling and simulation of a single-server queuing system [61]. This queuing system involves gating a fluidic sample into a chamber for a mixing operation within a larger protocol. The strategy for realizing this performance simulation capability by defining a set of data abstractions is proposed. In addition, services to augment a core simulation microkernel are also discussed. Experimental performance simulation results comparing the new composable component-based simulation strategy with a more traditional dedicated performance simulator are presented.

For the microfluidic mixing model, the interarrival time of fluidic samples, denoted by $T_{a_1}, T_{a_2}, \ldots$, are assumed to be independent, identically distributed (IID) random variables with a uniform probabilistic distribution. Without loss of generality, the lower and upper bounds of the uniform distribution are arbitrarily chosen to be 10 seconds and 19 seconds, respectively. A fluidic sample that arrives and finds the mixing chamber (service) idle enters service immediately. Due to the variability in fluidic samples (amount and composition), the mixing times of individual fluidic samples, denoted by $T_{s_1}, T_{s_2}, \ldots$, are also assumed to be IID random variables. The mixing times are assumed to be independent of the interarrival times, and their probability distribution is also uniform, having lower and upper bounds of 11 seconds and 17 seconds, respectively. A fluidic sample that arrives and finds the mixing chamber busy is queued into a set of reservoirs. When the mixing process completes for a fluidic sample, another fluidic sample is gated (valved) into the chamber, using a first-in, first-out (FIFO) discipline.

Performance statistics for the arrival, possible queuing, service time, and departure of a fluidic sample are collected in a VHDL record defined below.

```
type FLUIDIC_SAMPLE;
type NODE_PTR is access FLUIDIC_SAMPLE;
type FLUIDIC_SAMPLE is
  record
    ARRIVAL_TIME : TIME;
    SERVICE_TIME : TIME;
    LEAVING_TIME : TIME;
    NXT          : NODE_PTR;
  end record;
```

A first-in-first-out queuing discipline is used, represented by a linked list of records using VHDL access types. The linked list is encapsulated within an entity or a concurrent procedure and provided as a high-level data abstraction for ease of modeling [62]. Figure 3.3 shows the general structure of the VHDL stochastic discrete-event performance model and the principal data abstractions/services. Interarrival times and service times are generated from a uniform probability distribution and implemented using process suspension/activation [43]. Statistical reporting is implemented via text files. In addition, Table 3.1 presents the results of simulating the operation of the stochastic MEFS for mixing fifty (50) fluidic samples. ( Simulations were conducted using the Synopsys VHDL simulator.) The VHDL simulation results were checked by developing a comparable model for the commercial performance simulator SLAM and equivalent simulation results were obtained.

```
package QUEUING_PKG is
   ...
end package QUEUING_PKG;

entity QUEUE_FIFO is
   ...
end entity QUEUE_FIFO;

use STD.TEXTIO.all;

use IEEE.MATH_REAL.all;
 -- Stochastic Discrete Event
 -- Simulation Services

use QUEUING_PKG.all;

entity MEFS_PROTOCOL is
   ...
end entity MEFS_PROTOCOL;
```

```
architecture MIXING of
                    MEFS_PROTOCOL is
   ....
begin
 ARRIVAL: process
   ...
 end process ARRIVAL;

 RESERVOIR: QUEUE_FIFO
                    port map (...);

 MIXING: process
   ...
 end process MIXING;

 STATISTICS: process
   ...
 end process STATISTICS;
end architecture MIXING;
```

**FIGURE 3.3**
**Structure of VHDL stochastic discrete-event performance model. There are four concurrent processes at the top-level architecture.**

#### 3.1.2.3 Continuous Time System Simulation

Continuous-time simulation evaluates the behavior of a system's state as a continuous or piecewise continuous function of time. Behavior can be defined directly by means of state equations or indirectly by means of differential equations. Though state equations explicitly represent system behavior over time, they are generally difficult to derive [63]. Differential equations use the derivatives of state variables, representing the relationship of the rate of change of state variables. Thus, continuous time simulation concerns evaluation methods to solve a defining set of simultaneous equations; the present focus is on ODAEs.

To solve differential equations based on the discrete-event simulation kernel, the continuous differential equations need to be transformed into finite difference equations and solved via numerical integration. For example, the Improved Euler or second-order Runge-Kutta numerical integration method [64] can be used for this purpose.

The Runge-Kutta numerical integration algorithm is a member of the class of Taylor series integration methods. It solves a differential equation by using a trapezoidal approximation to a Taylor series expansion of the desired function. Using the Fun-

**Table 3.1** Statistical Analysis of Stochastic Microfluidic Mixing VHDL Performance Simulation

| Statistical Analysis: Microelectrofluidic Mixing<br>Date : 02/08/99<br>Run Number 1 of 1<br>Current Simulation Time : 740 | | | | | |
|---|---|---|---|---|---|
| | Average | Variation | Min | Max | Observation |
| Time in Mixing | 23.5 | 6.923 | 11 | 37 | 50 |
| | Average Length | Variation | Min Length | Max Length | Average Wait Time |
| Queue | 0.63 | 0.631 | 0 | 2 | 9.26 |
| | Current Utilization | Efficiency | | Max Idle Time | Max Busy Time |
| Mixing Chamber | 1 | 0.96 | | 19 | 158 |

damental Theorem of Calculus, the differential equation

$$\frac{dy}{dt} = y' = f(t, y) \tag{3.4}$$

can be rewritten as a finite difference,

$$y'(t_n + \theta\Delta t) = \frac{y(t_{n+1}) - y(t_n)}{\Delta t}, \qquad \exists\theta, 0 < \theta < 1 \tag{3.5}$$

which in turn yields the following equation:

$$y(t_{n+1}) = y(t_n) + \Delta t f(t_n + \theta\Delta t,\ y(t_n + \theta\Delta t)). \tag{3.6}$$

The term $f(t_n+\theta\Delta t, y(t_n+\theta\Delta t))$ is called the *mean gradient* in the region $[t_n, t_{n+1}]$ and is denoted by $K^*$. The second-order Runge-Kutta integration method computes the mean gradient by the weighted mean of the gradients $K_1$ and $K_2$ at the two points

$$\begin{aligned} t_1 &= t_n \\ t_2 &= t_{n+p} = t_n + p\Delta t, \quad 0 < p \leq 1. \end{aligned}$$

The mean gradient becomes

$$K^* = \lambda_1 K_1 + \lambda_2 K_2 \tag{3.7}$$

with the constraints that

$$\begin{aligned} \lambda_1 + \lambda_2 &= 1 \\ \lambda_2 p &= \frac{1}{2} \end{aligned}$$

to ensure second-order precision. Normally, $\lambda_1 = \lambda_2 = 0.5$, and $p = 1$.

Thus, the difference equation in (3.6) becomes

$$\begin{aligned} y(t_{n+1}) &= y(t_n) + \Delta t K^* \\ y(t_{n+1}) &= y(t_n) + \Delta t(\lambda_1 K_1 + \lambda_2 K_2) \end{aligned} \tag{3.8}$$

with

$$\begin{aligned} K_1 &= f(t_n, y(t_n)) \\ K_2 &= f(t_{n+p}, y(t_n) + p\Delta t K_1) \end{aligned}$$

Integration precision increases with decreasing stepsize $\Delta t$. However, decreasing stepsize $\Delta t$ increases calculation complexity, and can decrease intersection accuracy. Thus, selecting an appropriate stepsize $\Delta t$ is important. To that end, precision can be measured as the difference between the integration result with stepsize $\Delta t$ and the integration result with half-stepsize $\Delta t/2$, i.e.

$$\Delta = |y^{\frac{\Delta t}{2}}(t_{n+1}) - y^{\Delta t}(t_{n+1})| \tag{3.9}$$

To illustrate the modeling and simulation of a continuous-time system using VHDL, consider a biological system of parasites/hosts, also called predators/prey [61]. There are several examples in nature of parasites that reproduce by infesting host animals and, in the process, kill the hosts. This relationship causes the host and parasite population sizes to fluctuate. When the parasite population grows, the host population declines. The decline in the number of host, in turn, causes a decline in the birth rate of parasites and, consequently, the population of the host begins to increase. This process oscillates indefinitely.

Let $H(t)$ and $P(t)$ respectively denote the population of hosts and parasites at time $t$. Also, let $r$ denote the growth rate of hosts (excess of birth rate over death rate from natural causes) in the absence of the parasites. Then, the overall rate of change of the host population is given by

$$\frac{dH}{dt} = rH(t) - C_1 H(t)P(t)$$

where, the death rate of hosts ($C_1$) is proportional to the product of the numbers of parasites and hosts, $H(t)P(t)$. Assuming the death rate of hosts is the birth rate of parasites, the overall rate of change of the parasite population is given by

$$\frac{dp}{dt} = C_1 H(t)P(t) - C_2 P(t)$$

where, $C_2$ is the natural death rate of the parasites. This set of differential equations is solved via the second-order Runge-Kutta method, and implemented using the general-purpose programming language capabilities of the VHDL process statement.

Figure 3.4 shows the general structure of the VHDL continuous time performance model and the principal data abstractions/services. The architecture is composed of a single process that solves the set of differential equations, using the Runge-Kutta method supplied via the NUMERICAL_INTG_PKG package. The inner loop solves the differential equations for a given time step per a desired accuracy. The outer loop increments the time step. Simulation time is advanced with integration time steps via a wait statement. Solutions are reported via text files. The simulation services provided via the VHDL NUMERICAL_INTG_PKG package and implemented as a software component enables a particular simulation capability which can be assembled by using a relatively light-weight base kernel.

```
package NUMERICAL_INTG_PKG is
  procedure RUNGE_KUTTA ( ... );
   ....
end package NUMERICAL_INTG_PKG;

use STD.TEXTIO.all;

use IEEE.MATH_REAL.all;
 -- Continuous Time DAE
 -- Simulation Services

use NUMERICAL_INTG_PKG.all;

entity PRED_PREY is
   ....
end entity PRED_PREY;
```

```
architecture CONFIG of PRED_PREY is
   ....
begin
 -- initialization

 DIFF_SOLVE: process
  STEP_NUM: while loop
   ....
   STEP_SIZE: while loop
     -- Compute new time point
     -- using step size
     RUNGE_KUTTA( ...... );
     -- Compute new time point
     -- using half step size
     RUNGE_KUTTA( ...... );
     -- Check difference
   end loop STEP_SIZE;
   -- Record integration values
   -- Increment step count
   ....
   wait for PERIOD;
  end loop STEP_NUM;
 end process DIFF_SOLVE;
end architecture CONFIG;
```

**FIGURE 3.4**
**Structure of a VHDL continuous-time performance model (The Runge-Kutta method is coded in a process).**

Figure 3.5 shows the result of continuous time simulation, using a variable step size within the range 0.025 to 0.25 units for 500 time units. The coupled oscillatory behavior is demonstrated.

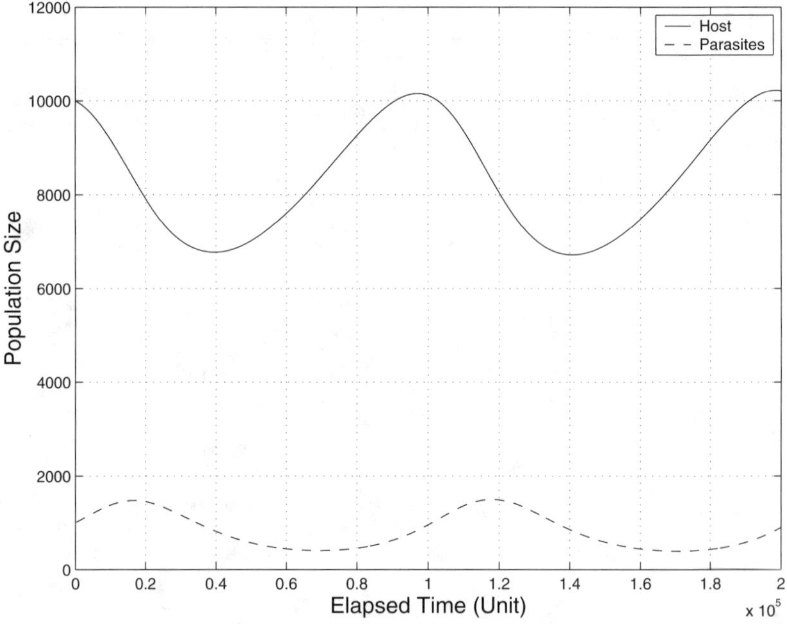


**FIGURE 3.5**
**Result of continuous time system simulation with VHDL. The population sizes of parasites and hosts are continuously changing with time.**

#### 3.1.2.4 Combined Discrete-Continuous System Simulation

Some systems exhibit both stochastic discrete and continuous behavior and thus, require a combination of the modeling and simulation strategies presented in the previous sections. The ability to model a combination of stochastic discrete and continuous system performance behavior with a single design capture and analysis strategy emphasizes the utility of an integrated performance analysis methodology.

Again, drawing from the domain of microfluidics, consider a biomedical application involving heating drug samples to enable subsequent chemical/biological analysis. The arrival time of drug samples is modeled as a random variable having a probability density function uniformly distributed between 2.5 minutes and 3.5 minutes. The initial temperature of each arriving drug sample is also a random variable that is uniformly distributed between $60^o$F and $80^o$F. The initial temperature of the heating process is taken as the average of the initial temperature of the drug sample and the ambient temperature that is assumed to be $300^o$F.

Let $O(t)$ and $U(t)$ respectively denote the temperature of the heating stage (oven) and drug sample (unit) at time $t$ and be defined by the following differential equations describing Newton's law of cooling. The time rate of change of the temperature of a body is proportional to its temperature and the temperature of the surrounding medium.

$$\frac{dO}{dt} = 0.12 \times (300 - O)$$

$$\frac{dU}{dt} = 0.04 \times (O - U)$$

A drug sample is removed from the heating stage when it reaches $250^o$F. Then, the drug sample at the head of the queue of reservoirs is selected for the next heating cycle.

There are two kinds of state variables in the described MEFS: discrete and continuous. The arrival of drug samples occurs at discrete points in time, whereas the temperature of a drug sample during the heating process changes continuously with time. The combined behavior is modeled using the representational strategies for stochastic queuing and numerical integration, along with the associated software components, as discussed in the previous sections.

Performance simulation results are shown in Figures 3.6 and 3.7. The operation of the MEFS is simulated for 60 drug samples, using a variable step size within the range of 0.4 to 4.0 seconds. The maximum single-step error is $1^o$F. Figure 3.6 shows the temperature of the ambient heating stage with respect to time. Solution of the differential and algebraic equations (DAEs) is shown using both Matlab and VHDL; the results are identical. Figure 3.7 shows the temperature of the drug sample with respect to time.

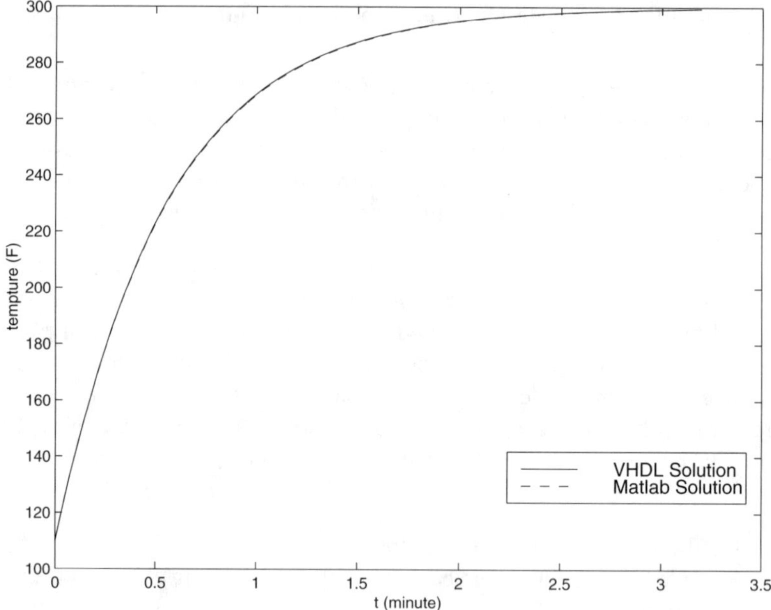


**FIGURE 3.6**
**Temperature change with time during heating. The VHDL solution and the Matlab differential equation solutions coincide.**

In summary, VHDL supports a broad range of constructs to describe detailed logic and abstract algorithms. It also provides both concurrent execution semantics and sequential execution semantics for parallelism and ordering procedures and functions. However, MEFS are object-oriented, and possess the queuing nature of higher-level stochastic behavior. The VHDL's event-driven perspective does not directly possess the capacity to represent this behavior. In addition, the language syntax prevents VHDL wider application from system-level modeling. Although the VHDL'92 version adds shared variables to enhance the VHDL's system-level design capability, the shared variable's constrained scope limits its application from the scalable system design objective, and there is still some controversy about the rationale of shared variables. Some research groups suggest extending the VHDL language syntax for system-level modeling [56], [65]. However, these extensions focus more on the electrical energy domain, and they can potentially destroy the integrity of VHDL. Hence, it is now well accepted that existing hardware description languages cannot be effectively expanded to support system-level modeling and simulation [66].

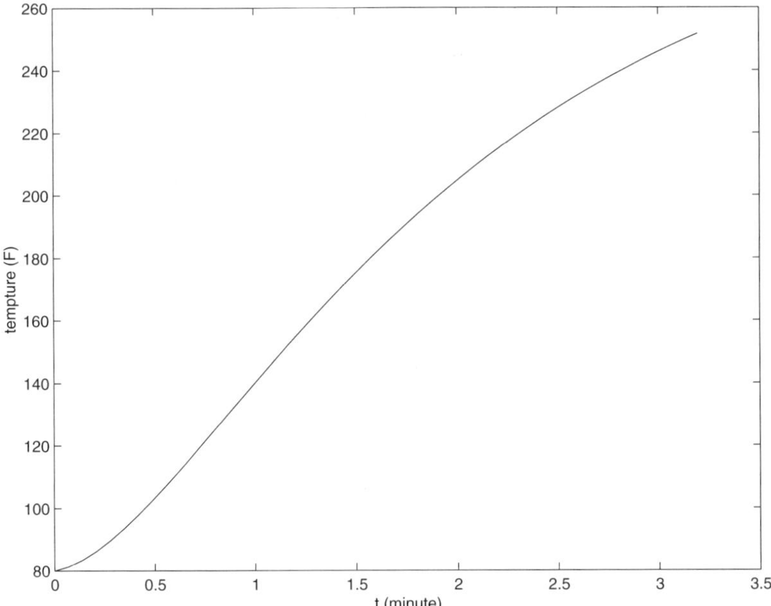


**FIGURE 3.7**
**Temperature change process of drug sample from $80^oF$ to $250^oF$**

### 3.1.3 Performance Language—SLAM

With the increasing complexity of MEFS system architecture, and its impact on system performance, it is becoming increasingly important to build a macro model for system architectural analysis. Such a model can improve computational efficiency, and lead to a better understanding of the system performance with different architectural configurations. Using an advanced simulation language to study system architecture performance can reduce model complexity, shorten model development time, and facilitate visualization of the system specification.

Advanced simulation languages, such as SIMAN [45], SLAM II [27] and SIMSCRIPT II.5 [26], are popularly used for system modeling and performance evaluation. These languages have the same basic modeling constructs due to language cross-fertilization over years of development. The first three languages are more suitable for queuing problems. SIMSCRIPT II.5 possesses more general process-oriented description capabilities, while it requires more lines of code for "standard" queuing problems. Table 3.2 shows the comparison between these simulation languages [67].

SLAM is an acronym for Simulation Language for Alternative Modeling. It is a FORTRAN-based simulation language. SLAM II is the latest release of SLAM, its newest PC version is Awesim [27]. It provides easy input procedures and output reports with hundreds of components. In addition, Awesim provides a more flexible system architectural description capability to develop a single simulation model. Moreover, Awesim provides sophisticated control statements and data structures to describe system behavior. Visual Basic and C++ can be coded in Awesim models. In particular, procedures are included to adaptively define files, entities, and collection variables. Alternative search procedures for locating entities are also provided. There are many new modeling capabilities in Awesim, and it is therefore more suitable for "complicated" simulation models. We next describe Awesim in more detail.

**Table 3.2** Comparison of Several General Purpose Simulation Languages

| Feature | GPSS/H | SIMAN | SIMSCRIPT II.5 | SLAM II |
|---|---|---|---|---|
| Event (E) or Process(P) | P | E.P. | E.P. | E.P. |
| Graphical model input | NO | YES | NO | YES |
| Combined discrete and continuous simulation | YES | YES | YES | YES |
| Standard Distribution Functions | Ex, N,T,U | Be,Er,Ex, Ga,L,N, P,T,U,W | Be,Bi,Er Ex,Ga,L, N,P,T,U,W | Be,Er,Ex Ga,L,N P,T,U,W |

| | | |
|---|---|---|
| Be: Beta | Bi: Binomial | Er: Erlang |
| Ex: Exponential | Ga: Gammar | L: Lognormal |
| N: Normal | P: Poisson | T: Triangular |
| U: Uniform | W: Weibull | |

### 3.1.3.1 Network Modeling

Awesim provides a process-oriented framework for modeling the flow of entities through processes. The framework is a network structure consisting of specialized *nodes* and *branches* that are used to model resources, queues for resources, activities, and entity flow decisions. An Awesim network model is a representation of a process and the flow of entities through the process.

To illustrate the network modeling of a system, a simple queuing system with a single-server is shown in Figure 3.8. Items arrive, wait, are processed by a single server, and then depart the system. There are three nodes and one activity. Creating node generates entities and routes them into the system. The interarrival time between entities is specified by a variable or a function. Queue node is a location in the network where entities wait for service. When an entity arrives at a queue node, it passes through the queue node and goes immediately into the service activity if the server is idle. If no server is available, the entity waits at the queue node. Queue node is a storage buffer with certain volume to store entities. Activities represent server activities. Terminate node is used to destroy or delete entities from the network. It can specify the number of entities to be processed on a simulation run. An entity can be assigned attribute values that enable a modeler to distinguish between individual entities of the same type or between entities of different type.

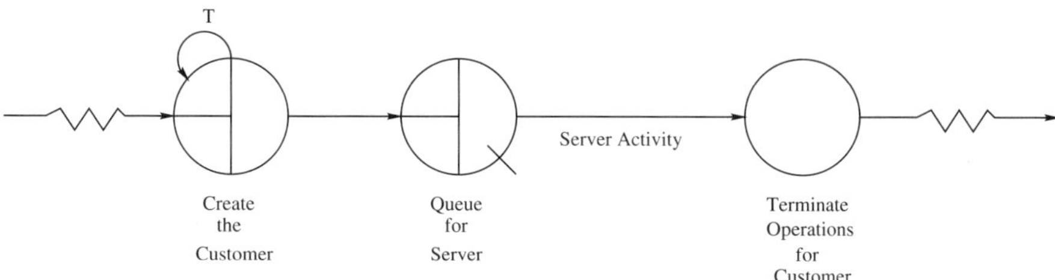


**FIGURE 3.8**
**A simple network model consists of a creation node, a queue node, a terminal node, and an activity branch.**

### 3.1.3.2 Event, Continuous and Combined Modeling

An important aspect of Awesim is that alternate worldviews can be combined within the same simulation model.

- Process Orientation
  The process orientation provides a concise and easy-to-learn modeling framework, but it lacks flexibility. As illustrated in Figure 3.8, Awesim employs a network structure to represent a framework model. This model pictorially represents the systems of interest, and the entities in the system flow through the network model.

- Event Orientation
  Event orientation provides difficult but highly flexible modeling framework. Awesim defines the events and the potential changes to the system when an event occurs. Mathematical and logical relationships are used to prescribe the changes associated with each event type. Awesim provides a set of standard subprograms to perform common discrete event functions, such as event scheduling, statistics collection, and random sample generation. Awesim provides time-advanced mechanisms to control the simulation. Additionally, Awesim

provides the initiating calls to the appropriate event subroutines at the proper points in simulated time. Hence, the modeler is completely relieved of the task of sequencing events to occur chronologically.

- Continuous Model
  Awesim codes the continuous model by specifying the differential or difference equations that describe the dynamic behavior of the state variables. These equations are coded by the modeler in Visual Basic and Visual C. When differential equations are included in the continuous model, they are automatically integrated by Awesim to calculate the values of the state variables within an accuracy prescribed by the modeler.

#### 3.1.3.3 User-Defined Function Interface

Awesim provides *nodes* to describe the entity flow to build the network model. In addition, Awesim supports user-written Visual Basic and Visual C inserts. The *EVENT* and *ENTER* nodes provide the key interface points between the network model and user codes. Awesim also provides the capability for allocating and freeing units of resources, altering resource capacities and specifying selection rules from user-written subprograms.

#### 3.1.3.4 Simulation Analysis

Awesim provides several standard distribution functions to modelers. These include the Beta, Erlang, Exponential, Gammar, Lognormal, Normal, Poisson, Triangular, Uniform and Weibull functions. Awesim also supports special user-defined functions. Moreover, Awesim provides several statistical reports for final data analysis. Awesim also provides simulation data for system performance evaluation. Examples of a system performance analysis are shown in Figure 3.9 and Figure 3.10. Figure 3.9 presents the detail simulation information for the number of jobs in the system when the $k$th job finished. Figure 3.10 shows the number of jobs in the system along the simulation time. Awesim also provides the animation to express the simulation results and the resulting status of the system.

In summary, SLAM is a high-level performance modeling language, which provides the capability to describe the overall system as a stochastic system, and provides a useful simulation methodology for performance evaluation. However, it lacks the capability to model and simulate hierarchical multiple-level MEFS behavior. Its modeling capability is limited to abstract high-level models, and it does not support component-level coupled-energy descriptions.

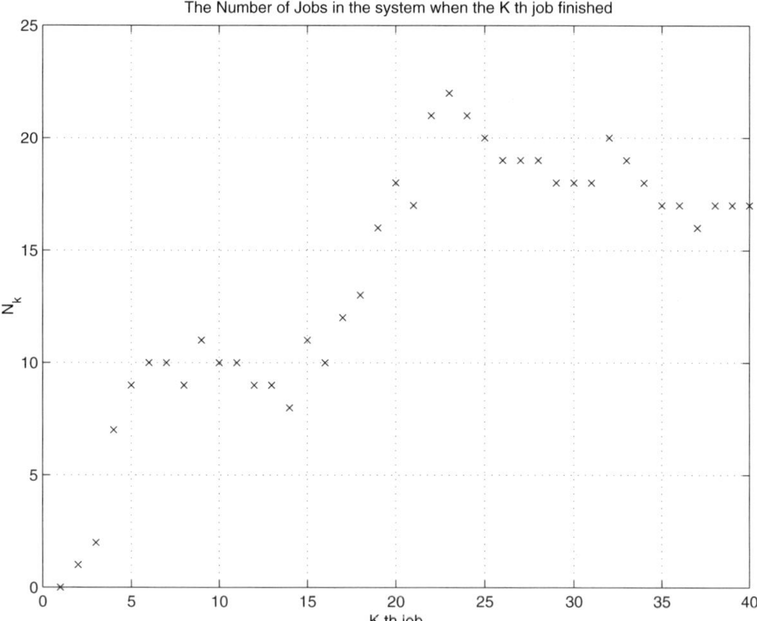


**FIGURE 3.9**
**The number of jobs in the system when the *K*th job finished.**

### 3.1.4 C/C++ and Matlab

#### 3.1.4.1 General Purpose Language - C/C++

C/C++ is a powerful and flexible language. In addition, C/C++ is a popular language preferred by professional programmers, and a wide variety of C/C++ compilers and helpful accessories are available. Moreover, C/C++ is a portable language. A C/C++ program written on a specific computer system can be compiled and run on another system with little or no modification. C/C++ provides very powerful dynamic data structures, such as pointers, linker and structure arrays. The flexible semantics and adequate mathematic functions make it possible to build any system model. C/C++ code can (and should) be written in routines called functions. These functions can be reused in other applications or programs. By passing pieces of information to the functions, the modeler can create useful, reusable code.

However, standard C/C++ does not possess the description capacity to directly study MEMS component-level coupled energy behavior. For example, C/C++ does not have a natural way to represent constrained data types, concurrency and clocks.

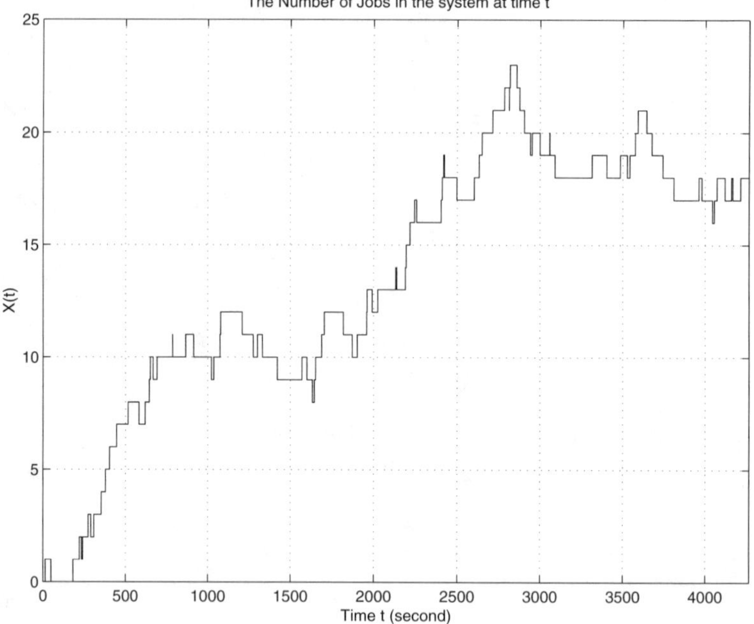


**FIGURE 3.10**
**The number of jobs in the system at time *t*.**

### 3.1.4.2 High Array Simulation Language - Matlab

Matlab is a powerful high-level language that is especially suitable for demonstrating mathematical concepts [68]. Matlab offers a useful working environment for model quick calculation and full simulation tasks. Additionally, Matlab, which is named as an array-based language, excels in the area of matrix computation; it has functions for nearly every type of matrix calculation. A variety of data types, such as matrices, data arrays, structures, character arrays, and cell arrays, can be created and loaded into Matlab. Moreover, the data can be analyzed and graphically visualized in numerous ways.

The Matlab C Math Library makes the mathematical core of Matlab available to application programmers. This library is a collection of approximately 300 mathematical routines written in C. Programs written in any other language capable of calling C functions can call these routines to perform mathematical computations. Matlab possess plentiful components and provides lots of powerful tools for special engineering calculation, which is very useful for system modeling and simulation.

Despite these advantages, Matlab lacks the capacity to describe the MEFS architecture. It does not support concurrent execution semantics and sequential execution semantics needed. In addition, Matlab does not possess the lower-level component modeling capability. For instance, Matlab does not support discrete event driven modeling, concurrent simulation, and multiple logical values. Moreover, there is general consensus among most Matlab users that certain Matlab programs run extremely slowly.

### 3.1.5 SystemC

Since C and C++ are the dominant programming languages, and C++ provides the capability to extend the language through classes without adding new syntactic constructs, a C++-based approach for hardware modeling is especially attractive. SystemC is a new open source library in C++. It supports the hardware-software codesign and the description of the architecture of complex systems consisting of both hardware and software components. It has been widely used in electronic hardware/software codesign [69], system-level design [70], and hardware synthesis [71]. SystemC and standard C++ development tools can be used to create a system model from the system level to the component level, quickly simulate to validate and optimize the design, explore various algorithms, and provide the hardware and software development team with an executable specification of the system. Based on the main features of SystemC version Beta v1.1 [72], the suitability of SystemC is discussed for building a hierarchical MEFS modeling and simulation environment.

#### 3.1.5.1 Module and Process

SystemC carries the notion of a container class called a *Module*. *Process* is used to describe functionality. This is a hierarchical entity that can have other *Modules* or *Processes* contained in it. *Module* and *Process* can have a functional interface, which allows us to hide implementation details and, in this fashion, include blocks of IP. In addition, *Process* can be stand-alone entities or can be contained inside a *Module*. SystemC provides *Module* and *Process* to describe the complex MEFS hierarchical architecture.

#### 3.1.5.2 Rich Set of Port, Signal, and Data Types

To support modeling at different levels of abstraction, from the system level to the component level, SystemC supports a rich set of port and signal types. They are very useful to describe the communication between different fluidic components. Additionally, SystemC supports a rich set of data types for describing the different fluidic sample properties, and multiple energy domains and abstraction levels. In contrast to VHDL/VHDL-AMS, which limit their variables' scope in single functional block, the scope of variables in SystemC is in the complete system. In addition, the fixed precision types allow fast simulation. The arbitrary precision types can be used for computations with large numbers and to model large buses. SystemC also includes a rich set of overloaded operators and type conversion mechanisms for those data types.

#### 3.1.5.3 Clocks and Reactivity

SystemC has the notion of clocks as special signals. Clocks are the timekeepers of the system during simulation. Multiple clocks, with arbitrary phase relationships, are supported. Besides the simulation clocks, SystemC includes an ultra light-weight cycle-based simulation kernel that allows high-speed simulation. Moreover, for modeling reactive behavior, SystemC provides mechanisms for waiting on clock edges, events, and signal transitions. SystemC also supports watching for a certain event, regardless of the execution stage of the process (the most common example is the watching of a reset signal).

#### 3.1.5.4 Enhanced Communication Protocols

SystemC provides multi-level communication semantics that enable designers to describe system I/O protocols at different levels of abstraction. In addition, SystemC supports a communication primitive called *channel.* A *channel* is a special type of signal that synchronous and asynchronous processes may use to communicate with each other. It can support complex communication protocols. Moreover, SystemC provides abstract ports that include communication semantics defined in the form of protocols.

#### 3.1.5.5 Multi-level Hierarchical Modeling and Simulation

The multiple-level abstraction design methodology is one of the most important properties of SystemC, ranging from the higher system level to the lower component level. SystemC provides the higher-level system specification to enhance MEFS top-down design process. In addition, SystemC's process-interaction worldview possesses the description capacity for MEFS higher-level queuing nature. Moreover, SystemC supports the event-scheduling representation, and the continuous perspective to describe the component behavior. To model and simulate continuous worldview with SystemC, differential equations with respect to time can be discretized and transformed into corresponding difference equations. Moreover, in contrast to the simulated event for the event-scheduling perspective of VHDL/Verilog, which means the simulation of program has to be connected with a simulator, SystemC provides the compiled event, which means the program can be compiled, then run independently of other simulators. The compiled event provides better simulation performance for large systems.

#### 3.1.5.6 Language and Associated Simulator

Due to the complexity of MEFS designs, it is necessary to relieve the system designer of the burden of simulator development. Designers mainly focus on the system modeling using related modeling and simulation languages. The associated simulator can automatically solve the system model with sophisticated mathematical methods, and it offers a flexible and standard interface for a user-defined program. Since SystemC does not provide an associated simulator, the designer is required not only to model the system behavior, but also to build the model solver.

**Table 3.3** Comparison between Different Simulation Languages

| Languages | SystemC | VHDL and VHDL-AMS | SLAM | Matlab | C/C++ |
|---|---|---|---|---|---|
| Associated Simulator | Poor | Good | Normal | Good | Poor |
| ODAEs Description Capability | Good | Normal | Poor | Poor | Good |
| Multi-level Description Capacity | Good | Poor | Poor | Poor | Good |
| Concurrency/Timing Mechanism | Good | Good | Good | Poor | Poor |
| Data Structure | Good | Poor | Normal | Normal | Good |
| Analysis Capability | Normal | Normal | Good | Good | Normal |

In summary, as shown in Table 3.3, after evaluating the suitability of these languages for MEFS hierarchical design, we conclude that SLAM II, VHDL/VHDL-AMS, C/C++, and Matlab are not suitable to handle the complete MEFS modeling and simulation. In contrast, SystemC is a viable candidate to develop a MEFS hierarchical modeling and simulation environment. This motivates the work reported in the rest of this chapter.

## 3.2 Building Design Environment with SystemC

In this section, a hierarchical modeling and simulation environment based on SystemC is presented. At first, in Section 3.2.1, the general characteristics of a hierarchical integrated design environment are discussed. Then, a MEFS system-level hierarchical modeling package with SystemC is presented in Section 3.2.2, and a MEFS circuit-level component modeling package with SystemC is discussed in Section 3.2.3. Finally, the numerical simulation package and the optimization/verification package are described in Section 3.2.4 and Section 3.2.5, respectively.

### 3.2.1 Hierarchical Design Environment

As mentioned in Chapter 1, on the analogy of the Gasjki and Kuhn's Y-chart in microelectronics CAD, as shown in Figure 3.11, a MEFS closed-loop integration design environment should extend the system design from the component level to the system level, and includes the following three functional blocks:

- Hierarchical Modeling and Simulation
- Hierarchical Design Optimization
- Hierarchical Design Verification

Therefore, the design environment consists of four different functional packages: system-level modeling package, circuit-level component modeling package, numerical simulation package, and optimization/verification package. These functional packages are discussed in the following sections.

### 3.2.2 System-level Modeling Package

System-level modeling involves the system performance modeling and the simulation of stochastic behavior in executing a specific biomedical and chemical application. In addition, system-level modeling studies the reconfigurable system architecture performance, scheduling, and throughput, etc. Therefore, it includes the following four functional blocks. As a special case study, a SystemC behavior model of a micro-chemical handling system is presented in Chapter 4.

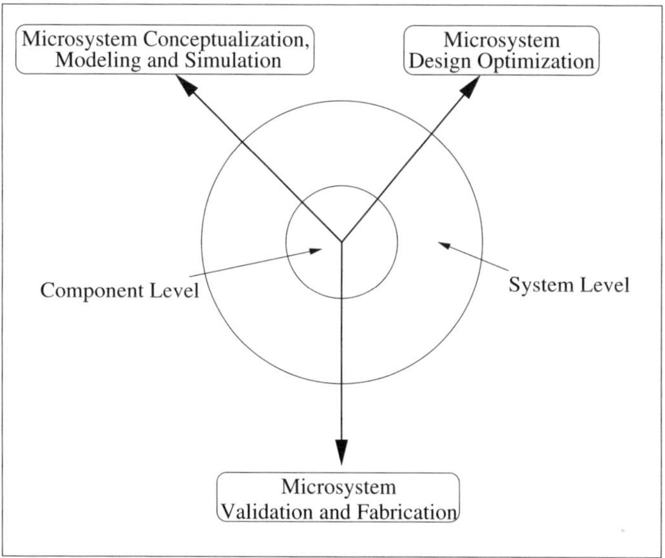


**FIGURE 3.11**
**MEFS closed-loop integrated design environment. It extends the system design from the component level to the system level, and includes three functional blocks.**

#### 3.2.2.1 Fundamental Elements

Depending on the system-level characteristics of MEFS, the fundamental elements for system modeling are:

- Storage part
  The storage part, such as a fluidic input buffer, is used to temporarily store the fluidic samples. When the fluidic sample enters the microfluidic system and the fluidic processor is busy, the fluidic sample is stored in the fluidic input buffer. Storage buffers are the independent functional blocks. The *module* construct can be used to model storage buffers. *Modules* include ports, constructors, data, and function members. In addition, *processes*, as the basic unit of execution, are used to emulate the behavior of the target device and system.

- Transportation part
  The transportation part, such as fluidic channels in the continuous flow system, or the two dimensional electrode arrays in the fluidic droplet movement system, is used to deliver fluidic samples from one site to another. The transportation time for each fluidic sample depends on the fluidic sample characteristics and the performance of actuators, such as micropumps or electrodes.

*Processes* can be used to model the functionality of the transportation part.

- Processor part
  Processors, such as fluidic analyzers, and mixers, etc., are the key parts for a MEFS bio/chemical application. Different processors have various processing properties. The processing time is based on the fluidic sample features and the processing function. *Module* is used to model the complete processor group. The individual processor functionality is defined with *Process*.

#### 3.2.2.2 Timing Clock

SystemC has the notion of clocks as special signals, which are the timekeepers of the system during simulation. In addition, clocks generate timing signals to synchronize simulation events. This allows parallel events to be properly modeled by the simulator on a sequential computer.

A clock object has a number of data members and methods to perform clock functionality. An example of a clock object is as follows.

```
sc_clock system_clock("System-clock", 20, 0.5, 1, false)
```

This declaration creates a clock object named "System-clock" with a period of 20 time units (The default time unit for SystemC is second), a duty cycle of 50%, the first edge occurring at 1 time units, and the first value being false.

#### 3.2.2.3 Fluidic Sample Declaration

MEFS system-level modeling is object-oriented, and it studies the change of fluidic sample characteristics. A complex and flexible fluidic sample data structure is built with SystemC. Table 3.4 shows the simulation result of a certain fluidic sample based on this data structure. It consists of the fluidic sample physical properties, and simulation procedure records. Each fluidic sample has a unique identification number. For instance, the ID of the fluidic sample shown in Table 3.4 is 12. The volume of the fluidic sample 12 is 30 units. Its processing purpose is for analyzing, and the analyzer 2 did it. At 56 time units, this fluidic sample enters the system (also called the generating time). At 70 time units, it arrives at the system storage buffer. At 107 time units, it arrives at the processor (analyzer 2). Total processing time is 30 time units, and the fluidic sample completes the processing at 137 time units. Fluidic sample 12 uses channel 1 for transportation from the inlet to the processor, and uses channel 2 for dispersing from the processor to outlet. The leaving system activity happens at 167 time units.

**Table 3.4** Fluid Sample Simulation Record

| | |
|---|---|
| Fluid sample: | |
| Fluid ID | = 12 |
| Volume | = 30 |
| ... | |
| Processing Purpose | = Analyzing |
| Processor ID | = Analyzer 2 |
| Generating Time | = 56 |
| ... | |
| Arriv. Buffer Time | = 70 |
| ... | |
| Pre Channel ID | = Channel 1 |
| Arriv. Processor Time | = 107 |
| Finish Process Time | = 137 |
| ... | |
| Post Channel ID | = Channel 2 |
| Terminate Time | = 167 |

#### 3.2.2.4 Fluidic Sample Transaction between Different Functional Blocks

The *Master* and *Slave* processes, which can perform data transactions based on an address, are used to perform fluidic sample transaction between different functional blocks. Using this mechanism, a *Master* process can write to or read from an address in a memory block. This memory block is in a *Slave* process. The syntax for *Master* and *Slave* processes is shown below

```
Syntax:
 sc_outmaster<fluid_type,
          sc_fullHandshake<fluid_type> > fluid_out;
 sc_inslave<fluid_type,
          sc_fullHandshake<fluid_type> > fluid_in;
```

where the "fluid_type" is the name of a fluidic sample structure. It consists of features of a fluidic sample shown in Table 3.4. The "fluid_out" and "fluid_in" are two variables possessing the "fluid_type" structure. Additionally, SystemC provides different communication protocols at different abstraction levels, from the abstract functional level to the detailed bus-cycle accurate (BCA) level.

### 3.2.3 Circuit-level Component Modeling Package

The goal of MEFS component modeling and simulation is to study individual microfluidic components at the circuit-level of abstraction, emphasizing the definition of physical properties and their relationships across multiple energy domains. Therefore, the circuit-level component modeling package includes the following four functional blocks. A special case studying for MEFS circuit-level modeling with SystemC is presented in Chapter 4.

#### 3.2.3.1 Energy Domain Behavior Declarations

Based on the microfluidic component modeling common characteristics discussed in Chapter 2, coupled energy component modeling requires the declaration of the common across variables and through variables to represent individual energy domains and disciplines. Therefore, by using *signal*, SystemC provides the declarations for the across and through variable for each energy domain. Figure 3.12 shows this declaration. They are grouped according to the energy domain.

```
-- ENERGY_SYSTEMS
signal <float> ENERGY;
signal <float> POWER;
signal <float> PERIODICITY;

-- MECHANICAL_SYSTEMS
signal <float> TRANSLATION;
signal <float> FORCE;
signal <float> ROTATION;
signal <float> TORQUE;

-- ELECTRICAL_SYSTEMS
signal <float> VOLTAGE;
signal <float> CURRENT;

-- FLUIDIC_SYSTEMS
signal <float> PRESSURE;
signal <float> FLOW_RATE;

-- THERMAL_SYSTEMS
signal <float> TEMPERATURE;
signal <float> HEAT_FLOW;
```

**FIGURE 3.12**
**The declaration of the common across variables and through variables for each energy domain using SystemC.**

#### 3.2.3.2 Coupled-energy Modeling

The special coupled-energy problems in MEFS require simultaneous statements describing concurrent events. These events can represent the dynamic behavior of components, processing events or transportation parts. Besides the standard C/C++ syntax, SystemC provides additional concept of *Process*, which includes three different types: *Methods*, *Threads*, and *Clocked Threads* to model the simultaneous activities in a system. *Process* is started or suspended when a certain condition is true, and returns control back to the calling mechanism when complete. The condition can be a clock edge, a variable, or a signal expression. As in VHDL, the concurrent processes in SystemC are loosely-coupled. The sensitivity list for each process has to be expressed explicitly.

#### 3.2.3.3 Conservative ODAEs Description Capability

As mentioned previously, lumped-element models are appropriate for describing the MEFS dynamic behavior with a conservative set of simultaneous ODAEs. These ODAEs governing composite microsystems possess a global structure reflecting the fact that physical systems obey laws of conservation of energy. In contrast to VHDL-AMS, SystemC does not directly provide constructs for defining sets of simultaneous ODAEs. Additionally, it is the user's responsibility to write and verify the energy-conservative models. However, its procedures provide powerful ODAEs description and solving capability.

#### 3.2.3.4 Analytical Modeling

An analytical model for MEFS has several advantages: rapid development (as compared to numerical model with FEM), adjustable parameters, and ease of system optimization. The analytical model of MEFS is described with the mathematical expression. SystemC provides a powerful and flexible description capability to build the analytical model within *Process*.

### 3.2.4 Numerical Simulation Package

In order to study MEFS stochastic behavior for a reconfigurable system architecture and a biomedical/chemical application, and to study the coupled-energy behavior of MEF components, a numerical simulation package is necessary. This numerical simulation package includes two parts: a real mathematical function, and mathematical solvers for DAEs.

#### 3.2.4.1 Mathematical Function

Some MEFS behavioral models can be solved analytically [16]. Therefore, a mathematical package with SystemC is required to provide the internal function to solve these analytical models. Moreover, MEFS system-level stochastic behavior requires this mathematical package contain the common real constants and common real probability functions. SystemC supports the capacity to build this mathematical package with the regular function procedures or *Process*.

#### 3.2.4.2 DAEs Solvers

MEFS component behaviors are associated with the ODAEs. Because these DAEs are coupled and some of them are inherently non-linear, they must be solved numerically and simultaneously with system simulation. Unlike VHDL-AMS, SystemC language does not directly provide an associated simulator to solve simultaneous ODAEs over a series of intervals denoting a period of time. Nevertheless, by using the regular function procedures or *Process*, users can code various DAEs solvers with SystemC, such as derivative and integral, and add them into a SystemC component behavior model. Moreover, besides the original simulation clock, SystemC can supply a higher-frequency clock to provide a series of time interval for more accurate ODAEs function solutions.

### 3.2.5 Optimization/Verification Package

With growing design complexity, fabrication process variation, and the harsh operating environment of MEFS, there is a need for hierarchical design optimization to support all aspects of product development, including design, manufacturing, and operational use. In addition, design optimization methodologies are required to match multiple design optimization objectives. Therefore, research is needed to investigate how to utilize the state-of-the-art optimization design methodologies in a hierarchical integrated design environment. As a part of integrated MEFS design environment, MEFS design optimization/verification includes two packages: a system-level optimization package, and a circuit-level optimization package. In combination with the state-of-the-art hierarchical design methodology, the detailed system-level hierarchical optimization and performance evaluation methodology is discussed in Chapter 4. The detailed circuit-level MEFS hierarchical design optimization/verification methodology is discussed in Chapter 5.

## 3.3 Conclusion

In this chapter, the suitability of several popular simulation languages is discussed for MEFS hierarchical modeling and simulation. Then, SystemC is demonstrated as a strong candidate for this purpose. The architecture of a hierarchical modeling and simulation environment based on SystemC consists of three tasks. Additionally, this design environment extends from lower-level component modeling and simulation to higher-level system modeling and simulation. Four functional packages have been described.

# Chapter 4

## System-level Simulation and Performance Evaluation

Current microelectrofluidic processors (e.g. biochips) have largely dedicated architectures supporting relatively specialized applications. A more general microelectrofluidic system computational architecture is required for performing a collection of differing analyses or procedures. However, the close integration of devices into this computational architecture is associated with strong energy-coupling issues. In order to support the growing complexity of MEFS design and to carry out global performance optimization, system-level performance analysis methodologies and tools are needed. These methods and tools should not only incorporate phenomenological laws from multiple disciplines, but they should also characterize dynamical behavior ranging from overall application execution to individual component operation.

In this chapter, we present the rationale, design, and simulation of next-generation microelectrofluidic system computational architectures for the emerging field of bioinformatics. In addition, we propose a hierarchical modeling and simulation methodology for MEFS. We also apply our architecture and methodology to the design of a micro-chemical handling system (MCHS). The modeling and simulation of MCHS is based on the SystemC design environment previously discussed.

The chapter is organized as follows. First, we presents a general microelectrofluidic system computational architecture involving a multi-drop bus, pipelined structure in Section 4.1. Next, the hierarchical modeling and simulation methodology is presented in Section 4.2. A micro-chemical handling system based on our general computational architecture is described in Section 4.3. Stochastic system-level and nodal component-level modeling based on SystemC are also presented. Hierarchical simulation results and system performance analysis optimization are then presented in Section 4.4. They address the system throughput, channel bus utilization, and reservoir capabilities. Finally, conclusions are presented in Section 4.5.

## 4.1 MEFS Computing and Architecture[1]

The purpose of the architectural study of a microelectrofluidic system is to develop design, analysis, and pilot implementation technologies for a reconfigurable microliquid handling system with biomedical applications. The goal here is to leverage component-level technology involving (e.g., pumps, valves, and reservoirs) to develop a microliquid handling system that can be reconfigured and reused for a variety of applications in biomedical miniature chemical analysis and precision fluid

[1]Reprinted from Proc. 3rd Int. Conf. Modeling and Simulation of Microsystems (MSM2000). A. Dewey, R. B. Fair, J. Jopling, J. Ding, T. Zhang, F. Cao, B. Schreiner, and M. Pollack, Towards Microelectrofluidic System (MEFS) Computing and Architecture, pp. 142-145, 2000. © 2000, with permission from Applied Computational Research Society (ACRS).

dispensing. The aim of analysis is to investigate a robust, hierarchical verification and optimization capability that encompasses both architectural system simulation with functional macro modeling, and circuit component simulation with lumped-parameter nodal modeling. Finally, pilot implementation investigates continuous and unit flow manufacturing strategies for cost-effectively constructing reconfigurable microliquid handling system prototypes. In Section 4.1.1, we discuss the general microelectrofluidic system architectural concepts. Next, based on the architectural proposal in Section 4.1.2, we present reconfigurable architectural functional requirements in Section 4.1.3. Then in Section 4.1.4, we propose a potential continuous-flow architecture. Unit-flow based architecture will be presented in Chapter 6. Section 4.1.5 discusses the system performance modeling and simulation methodology for microelectrofluidic systems.

### 4.1.1 Architectural Concepts

A microelectrofluidic architecture provides an integration framework for elemental computational components. Sample elemental computational components include:

| | |
|---|---|
| Pumps | Valves |
| Reservoirs | Chambers |
| Channels | Nozzles |
| Filters | Ports |
| Flow Sensors | Diagnostic Structures |

In developing initial architectural concepts for microelectrofluidics, it is instructive to attempt to "map" microelectrofluidic architectural concepts into microelectronic architectural concepts to gain insight into potentially useful organizational structures. For instance, reservoirs are similar to registers in that both components hold/store information across computational time-based boundaries. Channels are similar to wire networks in that both structures provide interconnection between components and the means to transport information. Valves are similar to gates in that both components provide selective connectivity to larger communication networks. Pumps are similar to voltage sources in that both components serve as power sources, activating components and moving information.

With this linkage, issues of optimal ways to sequence data (electric charge) movement through a computer to affect the execution of an instruction potentially apply to the movement of liquid through a microfluidic molecular system to affect the execution of a recipe. For example, issues of the design of power distribution for a computer apply to the design of pump apportionment for a microfluidic molecular system. Within such a conceptual framework, many of the organizational aspects of high performance register transfer data path and control flow computer architecture design can be profitably investigated for microfluidic molecular system architecture.

Conceptual electronic and fluidic architectural analogies are summarized in Table 4.1. Liquid is the "data" and is oriented primarily horizontally, whereas process control manages the sequencing of liquid movement and is oriented primarily vertically. Process control involving pump and valve activation provides the ability to enact several different fluidic-based process sequences (recipes) using the same canonical hardware. Acquisition (input) and dispensing (output) processing is decoupled by allocating separate reservoirs (register banks). Reservoirs are connected by a bus structure of channels and liquid storage. Similar to Arithmetic Logic Units (ALUs) or Memory Management Units (MMUs) for computers, various units that perform specific fluidic processing operations, such as agent detection or composition measurement, can also be connected to the channels. Fluidic analogies to electronic accelerator functions, such as floating point computation or network protocol encode/decode, are catalysts.

**Table 4.1** Microelectronic and Microfluidic Architecture Analogies

| Resource | Computer Architecture | Microelectrofluidic Architecture |
|---|---|---|
| Storage | Register | Reservoir |
| Transport | Wires | Channels |
| Effort | Voltage | Pressure |
| Flow | Current | Volume |
| Switching | Gates | Valves |
| Power Apportionment | Power Networks | Pressure Lines |
| Processing | Instruction Sequencing | Task Sequencing |
| Resource Utilization | Register Allocation | Reservoir Allocation |

### 4.1.2 Architecture Proposal

Building on the general discussion of microelectrofluidic architectural concepts discussed in the previous section, this section presents the structure of a preliminary microelectrofluidic architecture for reconfigurable computational microelectrofluidic systems. To understand the rationale for microfluidic system architectures, several key functional requirements and implementation constraints influencing the architecture design need be discussed. First, though the emphasis is on biomedical microliquid handling applications of agent detection and precision drug dispensing, the microelectrofluidic architecture should be able to support a wide array of microliquid handling processes involving sample and reagent acquisition, preparation, routing, transport, handling, and dispensing. These applications are typically continuous flow or analog operations, often involving highly custom microfluidic structures. This functional requirement implies a continuous flow "datapath" composed of a family of operations provided by a suite of multiple, possibly customizable, units. Different datapaths can be realized by instantiating different combinations of operational units. Conversely, control-oriented operations of input (acquisition) and output (dispensing) routing of fluid through the datapath and possible storage of intermediate results are often common across microliquid handling processes and, as such, represent opportunities for a common and consolidated "control path".

Another functional requirement involves the need to support multiple microfluidic recipes, with each recipe involving multiple steps. Multiple steps implies a level of component integration greater than two or three units. Multiple recipes implies the need for reconfigurability, meaning that flow networks connecting various microfluidic components can be changed to route different liquids through different sequences of operations. Additionally, reuse is important to avoid the limitations of one-time disposable units. The functional requirement of reuse implies system flushing strategies to prevent problems of contamination or errant chemical reactions. Finally, there is a need for throughput performance scalability, supporting a wide range of liquid volume flow rates with high precision.

Implementation constraints also have a significant influence on architecture design. For instance, microvalves are expensive and micropumps are even more expensive. Designing miniature analogs of macro valves in a manner conductive to microfabrication technology (lithography or micromolding) is presently a major technology challenge and an active research area. Present prototypes report moderate switching capabilities. Micropumps rely on microvalves for input and output rectifying flow and present additional energy consumption challenges related to realizing periodic pressure activation. Another implementation constraint concerns accommodating the wide array of microvalve and micropump actuation techniques, involving piezoelectric, thermopneumatic, electrostatic, electromagnetic, and bimetallic. Ideally, the microelectrofluidic architecture should not bias a particular manufacturing strategy.

### 4.1.3 Reconfigurable Architectural Functional Requirements

The much revered goal of such fludic systems is the much touted *laboratory-on-a-chip*, or *micro-total analysis system* ($\mu$TAS). Such a system must be versatile enough to handle a range of procedures for any laboratory or environment in which it would be employed, anywhere from a research laboratory to a crime scene to on-site emergency medical care to the front line of battle. This being the case, it is clear the potential applications are wide ranging.

This is where the use of analogies to the more developed electrical world are most useful. A detailed look at a set of potential applications reveals that despite the wide variety of procedures, there exists a fundamental 'instruction set' of fluidic and biochemical operations that can be 'reprogrammed' to accomplish countless larger scale operations (Figure 4.1). What ultimately makes each procedure unique is the final analysis step. In turn, each of these 'instructions' requires that certain hardware units be present for execution, much the way an ADD operation could not execute without an ALU.

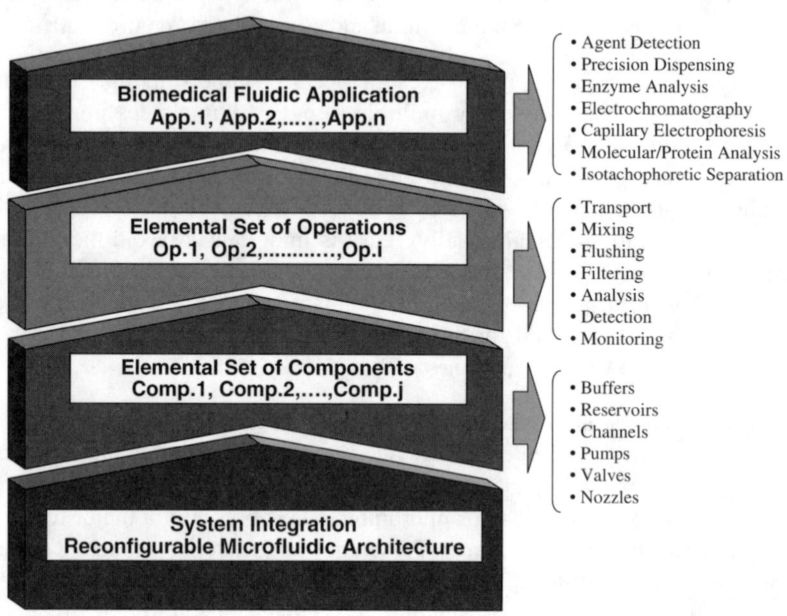


**FIGURE 4.1**
**Microelectrofluidic applications**

As mentioned above, the system must be fairly flexible in its programmability. This reconfigurability can occur on two levels, either by electrical control or through physical instantiation. Electrical control is little different from that found in established computational architectures, and is dictated by an overseeing program. Physical instantiation involves modular components that can be removed and added to the architecture to meet a specific demand.

### 4.1.4 Potential Architecture

Having identified the component set necessary to realize the instruction set, the question remains as to how to organize and interconnect them to give the best performance. Dealing in a fluidic/chemical domain rather than an electrical domain, the key is the observation that each canonical operation will likely have differing execution times. This can depend on the operation, the overall procedure performed, and the respective fluids involved. The simplest example is the execution of the final analysis step, which will typically last much longer than the preceding preparation steps. In addition, unlike in modern computer architectures, these operations cannot be further broken down into a series of small steps capable of being pipelined. The lowest level of pipeline granularity occurs at the level of the canonical operation.

This observation of disparate execution times implies a bus oriented architecture, where devices can be accessed in parallel with no requirement that one operation completes before another operation initiates. Furthermore, more frequently accessed and faster executing operations will necessitate more frequent use of the bus than comparably longer and infrequent operations.

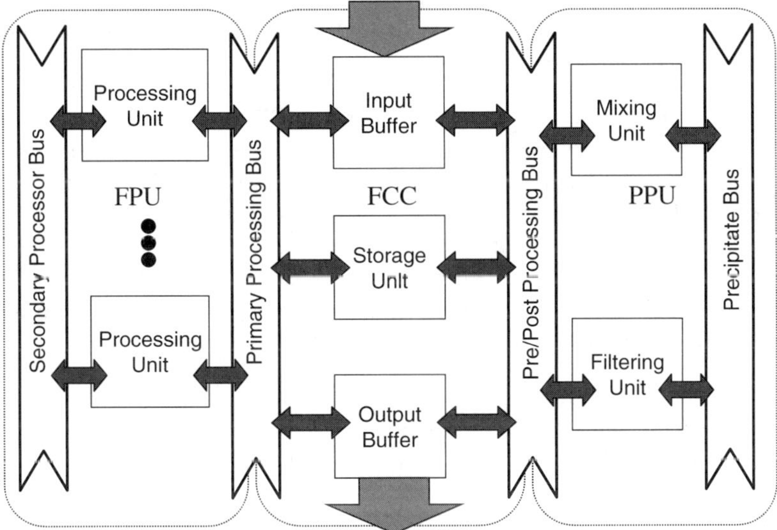


**FIGURE 4.2**
**Microfluidic architecture**

These factors are reflected in the architecture of Figure 4.2. The design is one centered around multiple buses. The *Fluidic Central Controller (FCC)* is the central storehouse, providing an interface into and out of the macroscopic world, as well as intermediate storage for running processes. The *Fluidic Processing Unit (FPU)* communicates with the FCC via the *Primary Processing Bus*. Units within the FPU may communicate amongst each other via the *Secondary Processing Bus* without having to occupy the primary bus. In addition, on the opposite side of the FCC, communicating via the *Pre/Post Processing Bus*, is the *Pre/Post Processing Unit (PPU)*, in which additional preparation and cleanup processes may occur. Similarly to the FPU, the units of the PPU may communicate with each other via the *Precipitate Bus*, where they may exchange reagents or expel waste by-products without having to occupy the Pre/Post Processing Bus.

### 4.1.5 Performance Modeling and Simulation

Figure 4.3 shows a Petri Net representation of the system architecture. The two major buses that connect the FCC, PPU, and FPU together are represented as resources. When a sample enters the system, it can take one of two branches, depending on which resources are available. For example, if the FPU bus is free, the sample will be transferred to storage units through the FPU bus. Once in the storage cell, it will again wait for the needed bus resource to be available. This time it is transferred to the PPU. This process will continue until the sample has finished all processing steps and is sent back to the storage cells in the FCC. There it waits for a bus to be available to be expelled from the chip.

A couple of notes about the model and simulation. First, it is necessary to observe that the peripheral buses are considered internal components of the FPU and PPU and, therefore, not explicitly in the Petri Net model. Second, this model assumes a simple procedure that requires only one trip each to the PPU and FPU to complete the operation. Input and output, of course, are the duty of the storage buffers in the FCC. This model also assumes that input fluids are first stored before processing occurs, and output fluids are stored before being expelled (as opposed to possibly buffering from input/output directly to/from a processing unit).

Figure 4.4 shows the system level schematic network modeling of this sequential Petri Net system, which benefits the study of system throughput, utilization, and the analysis of the bottleneck in the system performance.

Based on the Petri Net representation in Figure 4.3, we can build a system model and simulation to study the system performance. There are several fluidic and pre/post processing units in the architecture. Each of them processes just one unit of incoming fluid sample. Storage buffers include many cells of the same volume. There is only one bus between the FPU and FCC (the primary FPU bus), and one between the PPU and FCC (the PPU bus). The interarrival time of fluidic samples satisfies a certain probabilistic distribution.

Figure 4.5 shows throughput simulation results. The horizontal axis denotes the sample input rate. The vertical axis represents what proportion of the inputs are accepted and processed by the system. The straight line shows an ideal case when availability of the system is guaranteed and all input samples are accepted and processed. The other curve is actual system performance. When input rate is low, throughput is nearly linear – the performance of the system approximates the ideal when input rate is low. At increased input rates, actual throughput drops below the ideal and quickly reaches saturation. At saturation, throughput remains constant regardless of input rate variation. Saturation is a common phenomenon in the study of system perfor-

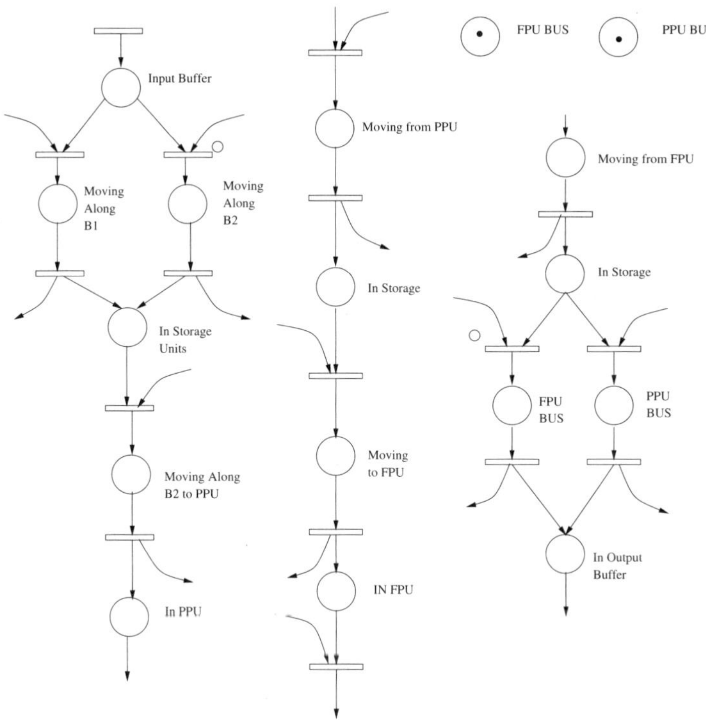


**FIGURE 4.3**
**Petri Net representation of architecture**

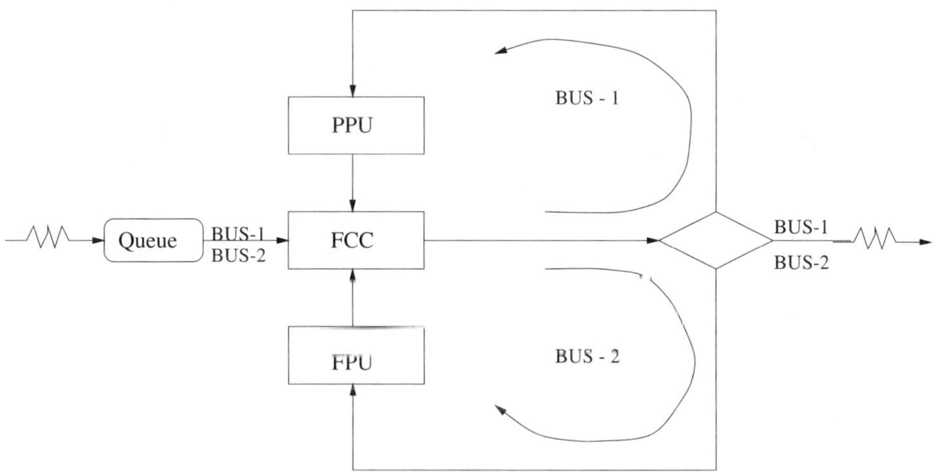


**FIGURE 4.4**
**Schematic network model of architecture**

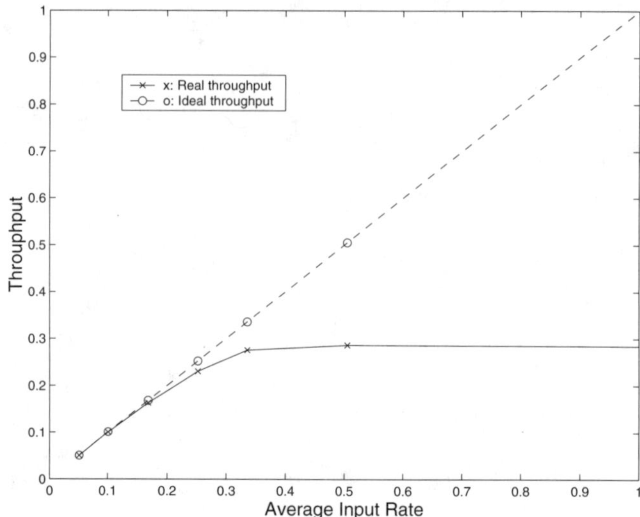


**FIGURE 4.5**
**Actual vs. ideal system throughput**

mance – it tells us that the resource is limited with respect to demand. The closer to the saturation point the system is operated, and the higher the saturation level of an architecture, the more efficiently we use available resources.

System architecture optimization is the key for higher system performance. Figure 4.6 shows the bottleneck of the system. The horizontal axis shows processing time normalized by bus transfer time. The vertical axis represents utilization. At low execution time ratios, transferring samples through the bus takes longer than processing samples. Therefore, the system bus is heavily utilized and processing units are under utilized (utilization below 0.1). There is a bottleneck at the point of communication. On the other hand, when bus transfer time is $\frac{1}{4}$ of processing time, processing becomes the bottleneck. More processing units would be needed to keep the system running at full speed. The ideal case for this model is when bus transfer time is exactly $\frac{1}{3}$ of processing time, where both communication resources and processing resources are well utilized.

Furthermore, the hardware design of microsystems is also related to the control software design. Figure 4.7 shows the utilization of each functional unit with a certain fluidic processing route control schedule. Although either bus (primary FPU or PPU) may become the bottleneck to improving the system performance, the FCC is the center of MEF systems. Its architecture and control scheduling should be carefully designed, otherwise "deadlock" due to cycle resource requirements between the related functional units may happen to make the system collapse.

MEFs has advanced immensely in recent years. However, designs and devices are

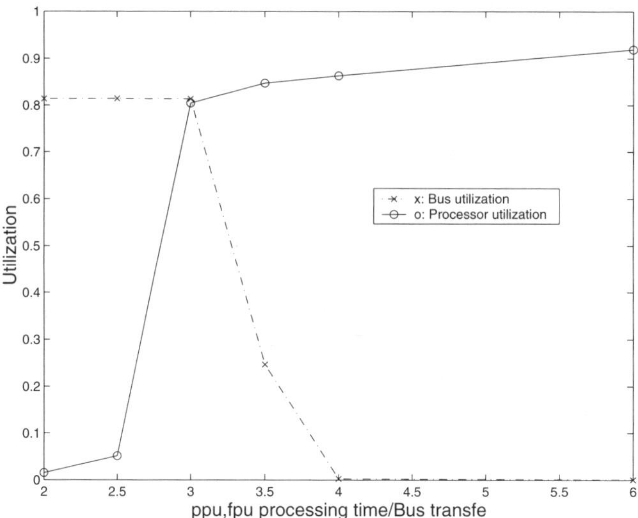


**FIGURE 4.6**
**Bottleneck of system architecture**

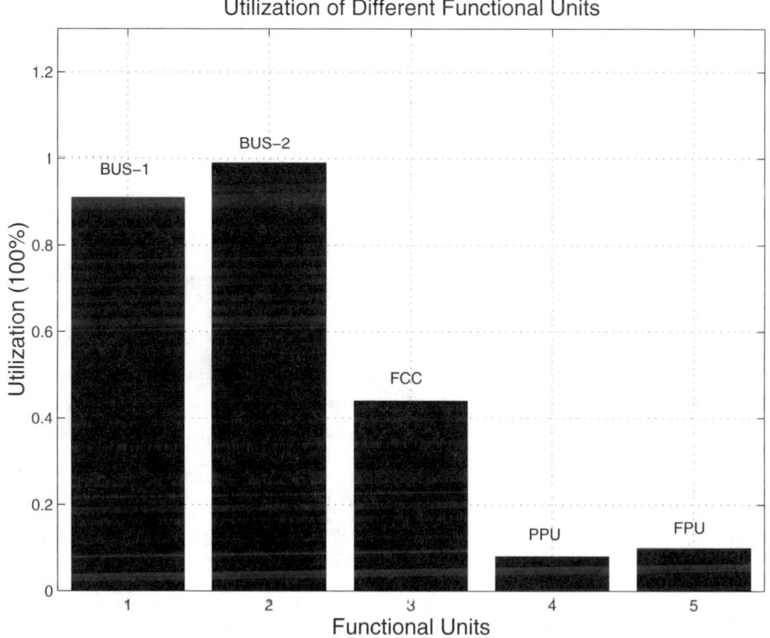


**FIGURE 4.7**
**Utilization of different functional units**

still largely application and function specific. Here, the design of a larger reconfigurable MEFS architecture has evolved from the demands that executing many application sets require. The system has logical functional units (FCC, FPU and PPU) and an interconnecting bus system that allows communication internally.

Performance simulation of such an architecture indicates that optimum performance is highly dependent upon the relationship between transit time and actual processing time. In addition, the distribution of process times associated within one functional unit (FPU or PPU) will play a role in optimizing bus and system usage. Furthermore, control scheduling is observed to also be a potential point for throughput bottlenecking or system dead-locking. Armed with this new knowledge, further efforts plan on implementing physical models, exploring these relationships and seeking out other potential hazards and places for improvement. These discoveries are then fed back onto the design of the architecture to successively move closer to the design of a true laboratory on a chip.

## 4.2 Hierarchical Modeling and Simulation Methodology[2]

In this section, we introduce the main concepts underlying hierarchical modeling and simulation for MEFS. These concepts are related to the MEFS hierarchy, the hierarchical modeling and simulation strategy, and software implementation issues.

### 4.2.1 MEFS Hierarchical Perspective

Figure 4.8 illustrates the hierarchy inherent in MEFS. Each layer of abstraction presents unique model fidelity, domain representation, and simulation efficiency requirements and challenges. Stochastic modeling and simulation provides a level of abstraction for studying architectural performance issues such as the architectural bottlenecks and capacity. Process flow modeling and simulation provides an additional level of abstraction for studying biomedical application execution issues, such as the throughput and overall execution times. Circuit-level component modeling and simulation provides a level of abstraction for studying the microfluidic component behavior in more detail. The ultimate objective of component modeling is to find a common modeling style and analysis strategy for the electrical, mechanical,

[2]This section is based in part on "T. Zhang, F. Cao, A. Dewey, R. B. Fair and K. Chakrabarty, Performance analysis for microelectrofluidic system using hierarchical modeling and simulation. IEEE Transactions on Circuit and System II: Analog and Digital Signal Processing, vol. 48, no. 5, pp. 482-491, May 2001." 

and fluidic domains involved in MEFS. In addition, component simulation models are based on the ODAEs underlying these domains. They are also based on a deeper understanding of the physical principles governing the microfluidic device behaviors.

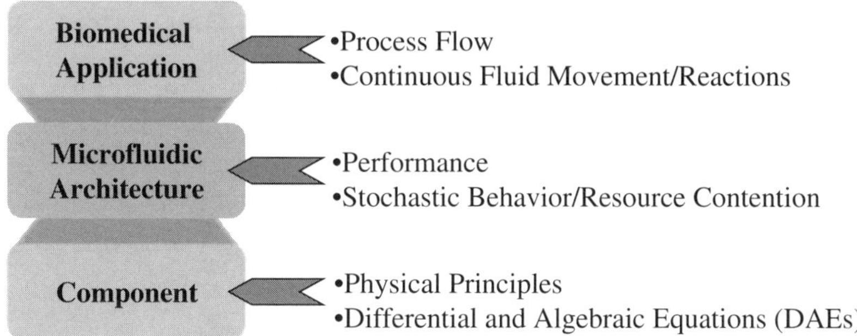


**FIGURE 4.8**
**Integrated modeling and simulation hierarchy for microelectrofluidic systems (MEFS) consists of three levels of abstraction.**

### 4.2.2 Hierarchical Performance Evaluation Strategy

MEFS performance analysis is difficult because coupled-energy behavior creates strong links between high-level architecture and low-level component design parameters. For instance, micropump actuation frequency influences fluidic sample transport rates, which, in turn, influences overall reaction and dispensing rates [4]. As another example, reservoir allocation, reaction scheduling, and time sequencing of microvalve settings influence resource utilization and overall area requirements. Therefore, a new MEFS hierarchical design strategy is required that incorporates phenomenological laws from multiple disciplines to characterize dynamical behavior ranging from overall application execution to individual component operation.

We adopt the strategy of trading-off behavioral fidelity with the efficiency of analysis, "blinding" unnecessary low-level detail, and paying more attention to certain tractable subsystems [73]. This encompasses architectural system simulation with stochastic macro modeling, and circuit component simulation with lumped-parameter nodal modeling. A hierarchical modeling and simulation methodology, shown in Figure 4.9, can then be developed to deal with the performance analysis challenges of MEFS. Each level of the modeling and simulation hierarchy possesses a unique set of representational conventions and simulation methodologies that are, for designer convenience, provided by a set of data and operator definitions. Different functional blocks are refined to different level based on the design requirements. Some of them, such as $A$ and $B$, are refined to the deeper level for providing more detailed design information. Some of them, such as $C$ and $D$, just provide abstract information.

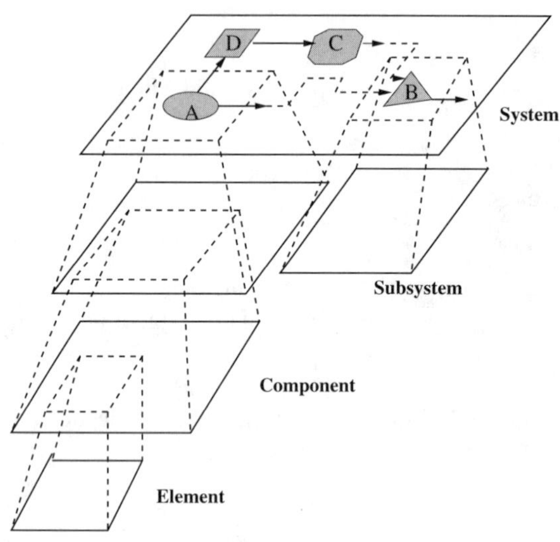


**FIGURE 4.9**
**Schematic view of hierarchical modeling and simulation; different functional blocks are refined to a different level based on the design requirements.**

### 4.2.3 Modeling and Simulation Language

Traditionally, different modeling and simulation languages are used for describing the unique set of representational conventions and simulation methodologies for each level of hierarchy. However, this multidisciplinary system design requires human interaction. It also leads to problems of misinterpretation of concept specifications in the translation between different data models and tools. Therefore, the SystemC hierarchical design environment discussed in Section 3.2 is used for complete MEFS hierarchical modeling and simulation.

## 4.3 Micro-Chemical Handling System

With the MEFS device capabilities advancing and discrete components improving[17], [16], small systems that combine existing components into a more useful device have all been conceived and many have been built, for example, devices for DNA analysis [2], chemical analysis [74], and more. The micro-chemical handling system (MCHS) is a well-known promising microsystem technology application in the chemical analysis area, which provides multiple processing functions, such as liquid mixing, analyzing and catalyzing, on a single microsystem. Based on the reconfigurable microliquid handling system architectural design discussed in Section 4.1 and the processing elements structure [75], the potential architecture of a micro-chemical handling system is presented in Figure 4.10.

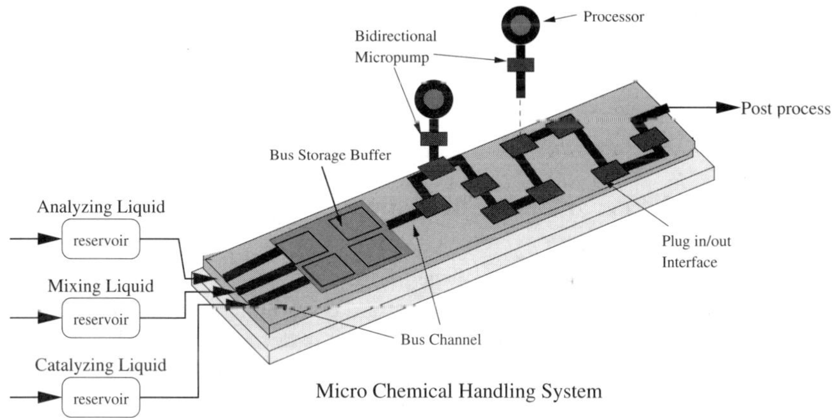


**FIGURE 4.10**
**Micro-chemical handling system consists of processing elements and a reconfigurable mother-board.**

The architecture of MCHS is composed of processing elements and the reconfigurable mother-board. The reconfigurable mother-board contains the liquid entrances, a bus storage buffer containing $n$ equal volume cells, and a single bus channel connecting the bus storage buffer to the exit. A total of $M$ standard I/O interface ports are located along the bus channel to connect the processing elements. A standard processing element has a standard I/O interface port, a bidirectional micropump providing pressure-driven flow throughout the system, and a reaction chamber involving mixers, analyzers, and catalyzers. Liquid samples enter the system from three containment reservoirs. Samples entering the system are queued in a bus storage buffer.

Micropumps intermittently draw liquid samples from the bus storage buffer to individual processors when the bus channel and related processor is available. After processing, the sample is pumped from the processor to the output, again when the bus channel is available. Due to the time spent transporting a liquid sample into a processing element, executing the chemical reaction, and transporting the resulting sample out of the system, chemical liquid samples initially input into a bus storage buffer until applicable resources are available. Due to the limitation of small displacement of the microvalves, sample solutions are filtered. Also, processors and the bus channel are periodically flushed with filtered calibrant solution between protocols.

System performance is related to fluid flow rate, liquid arrival traffic, the structure of the channel bus, the number of different kinds of processing elements, fluidic sample processing time, reservoir capabilities and processing scheduling, etc. The micropump is one of the main components of a micro-chemical handling system; it provides the liquid driving power in the channel and processors. It determines the important characteristic of fluid flow rate. In addition, the special chemical reaction functionality is also a key to system performance. The following sections show how the performance of the MCHS can be evaluated under variations of the system design parameters.

### 4.3.1 Stochastic Performance Modeling

The architectural simulation model of the micro-chemical handling system is developed using a combination of process, event, and continuous control paradigms. The operation of the micro-chemical handling system is denoted by the flow of entities (clients) through a network structure consisting of nodes and branches denoting resources, queues for resources, activities, and entity flow decisions.

Figure 4.11 shows a diagram of the queuing network model of the micro-chemical handling system. This model is built based on the SystemC design environment. The chemical handling process can be separated into five stages, depending on the liquid-process routine. The first stage—the initial sample creation, is coded in *producer.h*. The second stage is initial sample acquisition—fluidic sample entering and being stored in an appropriate containment reservoir. This stage is coded with *reservoir.h*. In stage three, a fluidic sample is moved from containment reservoirs into the bus storage buffer that is not full. The behavior of this stage is coded in *busfifo.h*, Then when the channel bus and a related processor are free, the fluidic sample is transported to the appropriate processor. The procedure *processor.h* is used for this stage. It consists of all the processors such as mixers, analyzers, and catalyzers. Figure 4.12 shows its header and implementation code of one analyzer. After processing, the processed liquid sample is moved out from the outlet when the bus channel is available. This terminal stage is coded in *terminal.h*. Simulation results

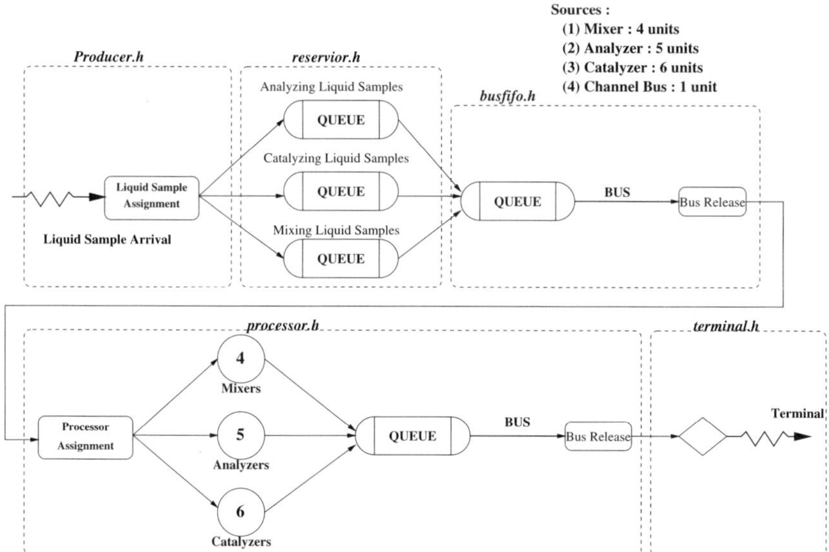


**FIGURE 4.11**
**A stochastic network model of the micro-chemical handling system includes five stages.**

are also recorded in this stage for further data analyses.

Figure 4.13 shows the program structure of this micro-chemical handling system. Each functional block is hierarchically connected to the higher-level program. The connection between different functional blocks is defined on the higher level. Associated numerical simulation packages and optimization packages support the system modeling, simulation and optimization. Figure 4.14 shows the MCHS top-level program structure based on SystemC.

Without loss of generality, the volume for each liquid sample is assumed to be the same ($800\mu l$) and equal to the cell volume. The three containment reservoirs provide an interface between the synchronous micro-chemical handling system and its asynchronous macro environment. The acquisition $f(x)$ as a function of time $x$ is modeled by a traffic of liquid samples separated by interarrival times, denoted by $\{T_1, T_2, ...\}$. They are independent, identically distributed (IID) random variables, and are characterized by an exponential probabilistic distribution given by (4.1) having a mean value of 15 seconds, that is $\lambda = \frac{1}{15}$.

$$f(x) = \lambda e^{-\lambda x} \qquad x \geq 0 \tag{4.1}$$

An input liquid sample enters the bus storage buffer immediately if the related processor (one per sample) is idle and the bus storage is not full; otherwise, the input

```
// header file: processor.h
SC_MODULE (processor)
{// define interface
  sc_outmaster<int> fluid_select;
  sc_inmaster<fluid_type > fluid_in;
  sc_outmaster<fluid_type> fluid_out;
  sc_in_clk clk;
  processor_buffer<fluid_type>
     Analyzer, Mixer, Reactor, ...;
  void analyzer();
  void mixer();
  void reactor();
  SC_CTOR(processor)
  {
  SC_THREAD(analyzer);
   sensitive << clk;
   SC_THREAD(mixer);
   sensitive << clk;
  SC_THREAD(reactor);
  sensitive << clk;
  }
};
```

```
// implementation file:analyzer.cpp
void processor:analzyer()
 {
 while (true)
  {wait();
   fluid_select=analyzing;
   // occupy the channel
   ...
   item=fluid_in;
   //delivery
   ...
   //processing
   ...
   //delivery
   ...
   fluid_out = item;
   //release the analyzer
   ...
   //release the channel
   ...
  }
 }
```

**FIGURE 4.12**
**Program structure for processor.h and analyzer.cpp.**

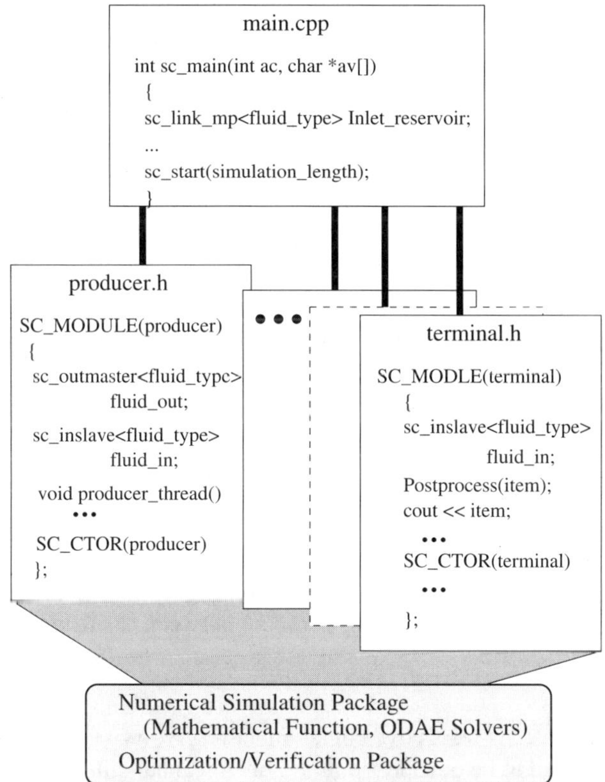


**FIGURE 4.13**
**Description of the model of a micro-chemical handling system based on SystemC. Each functional block is hierarchically connected to the higher-level program.**

```
#include "systemc.h"

#include "fluid.h"
-- define fluidic features.

#include "config.h"
-- define system architecture.

#include "producer.h"
-- generate fluidic samples.

#include "busfifo.h"
-- define storage buffer

#include "reservoir.h"
-- define reservoir config.

#include "channel.h"
-- define channel config.

#include "processor.h"
-- define process elements
...
```

```
int sc_main(int ac, char *av[])
 {
 //Signals
 sc_signal <int> bus_storage;
     ....
 sc_link_mp<int> reservoir_select;
 sc_link_mp<fluid_type>
                  producer_reservoir;
     ....
 sc_clock genClk("genClk",
                  1, 0.5, 0.1, true);
     ....
 // instantiate functional blocks.
 producer fluid_producer("Master");
     ....
 reservoir reservoir("Reservoir");
     ....
 busFifo storage("StorageBuffer");
     ....
 processor processors("Processor");
     ....
 terminal Terminal("TerminalNode");
     ....
 sc_start(simulation_length);
 return 0;
 }
```

**FIGURE 4.14**
**Top-level structure of a micro-chemical handling system model based on SystemC. It defines the communication protocol between the functional blocks.**

liquid is queued into its respective containment reservoirs. The containment reservoirs are considered to be outside the micro-chemical handling system and they use a first-in, first-out (FIFO) queuing discipline.

Depending on the basic architectural organization and execution concepts for assembling various microfluidic devices into a network [76, 77], three general kinds of chemical reactions are modeled to support a diverse set of applications: analyzing, mixing, and catalyzing. The processing time is based on the fluidic sample properties and processing functionality. Here we assume the analyzing time and mixing time are fixed as 20 seconds and 50 seconds for each fluidic sample, respectively. The special catalyzing functionality is discussed in the next section. There are 3 analyzers, 4 mixers, and 2 catalyzers. Forty percent of the fluidic samples require analyzing, 30 percent of the fluidic samples require mixing, and 30 percent of the fluidic samples require catalyzing. There is one channel bus. When the channel bus is free, a liquid sample waiting in the bus storage buffer can be transported to its associated processor when that processor is free. A first-request-first-occupy priority protocol is used for channel assignment. Delivery time depends on the distance from the bus storage buffer to the particular processor, modeled by a geometric distribution. The total length of the microchannel $L$ is separated into several equal segments based on $M$ that is the number of standard I/O interface ports locating along the bus channel. Each processor is located at each interface port sequentially. Therefore, the distance from the bus storage buffer to analyzers, mixers, and catalyzers are $\frac{i}{M}L\ (i = 1, 2, 3)$, $\frac{i+3}{M}L\ (i = 1, 2, 3, 4)$ and $\frac{i+7}{M}L\ (i = 1, 2)$, respectively. The total number of processing elements is less than or equal to $M$. Here, we assume that $M$ is equal to 10, and the length of the channel bus is 12mm. After processing, liquid samples are transported out of the micro-chemical handling system.

### 4.3.2 Thermal Catalyzing Process Functionality

Similar to the heating process discussed in Section 3.1.2.4, the catalyzing process is a fluidic-sample thermal-reaction process. Its functionality is very useful for various bio/chemical analyses, such as biochemical reactions with DNA [2]. After the fluidic sample arrives at the reaction chamber, it is heated. When the fluidic sample reaches a certain temperature, the catalyzing process finishes, and the fluidic sample is pumped out.

Again, we assume that the initial temperature of each arriving fluidic sample is a normally-distributed variable with mean $\mu = 70^o$F, and the variance $\sigma^2$ as $0.8$. The initial temperature of the thermal reaction chamber is taken as the average of the initial temperature of the fluidic sample and the initial chamber temperature assumed to be $300^o$F. Let $y(t)$ and $x(t)$ respectively denote the temperature of the heating stage (chamber) and the fluidic sample (unit) at time $t$. The temperature-changing rate of a fluidic sample is proportional to its temperature and the temperature of the surrounding medium.

$$\frac{dy(t)}{dt} = 0.12 \times (300 - y(t)) \tag{4.2}$$

$$\frac{dx(t)}{dt} = 0.04 \times (y(t) - x(t)) \tag{4.3}$$

When a fluidic sample is removed from the reaction chamber after it reaches $250^o$F, another fluidic sample at the head of the queue of the storage buffer is selected for the next thermal catalyzing cycle. There are several numerical integration methods that can be used to solve the ODAEs (4.2) and (4.3). Here, relaxation-based numerical integration techniques [64] coded in SystemC are used to solve these ODAEs. In order to improve the accuracy of the computed results, another simulation clock with higher-frequency is used. Figure 4.15 shows the integration result of a certain fluidic sample with the initial temperature at $54^o$F.

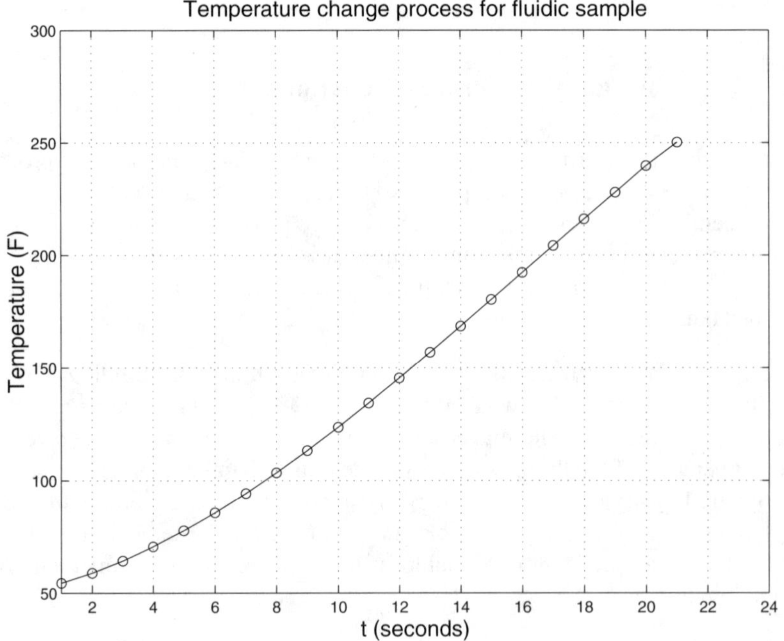


**FIGURE 4.15**
**Temperature change process of a fluidic sample from $54^o$F to $250^o$F**

### 4.3.3 Microvalve Lumped-element Nodal Modeling

Micropumps are the major components of microfluidic process system. Usually, they consist of an actuation unit and two passive check valves, and microchannels are used for connecting the inlet part and outlet part. The pressure-driven check valves are very important to the behavior of the micropump since they determine the flow rate of the micropump. The major parts of the check valve are a cantilever beam and valve seats. Normally the cantilever lies against the valve seat, thus closing the port to fluid flow. In operation, the fluid flow will exert pressure against the cantilever. The cantilever, acting like a spring, will deflect and allow the fluid flowing through the valve. The schematic view of a valve is shown in Figure 4.16 [16].

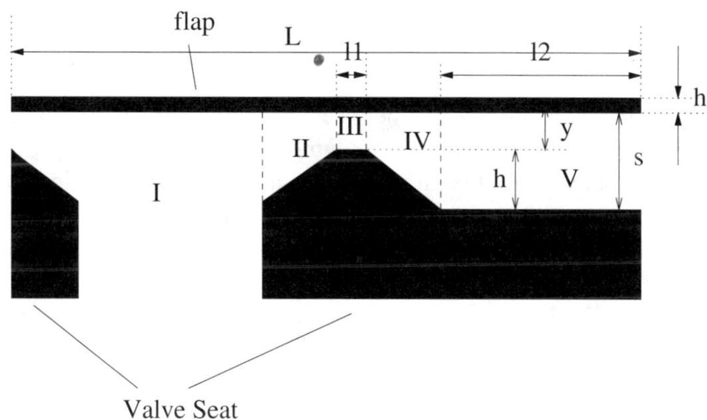


**FIGURE 4.16**
**Schematic view of the opening valve. The gap between the cantilever and the valve seat is divided into five pieces.**

Static flow rate is an important design consideration, and it is dependent on the structural parameters and the displacement of the valve. The displacement is determined by the pressure difference between the ports

$$\Phi = f(y, y_1, y_2, ...y_n) = g(y_1, y_2, ..., y_n, p) \tag{4.4}$$

where $\Phi$ is the static flow rate, $y$ is the displacement, and $y_1, y_2, ..., y_n$ stand for the structure parameters, and $p$ is the pressure difference between the inlet and the outlet.

The behavior of the cantilever can be described by a second-order differential equation,

$$m\ddot{y} + d\dot{y} + ky = pA \tag{4.5}$$

where $m$ is effective mass of the cantilever, including the mass of the cantilever and that of the liquid surrounding the cantilever. The parameter $d$ is the damping constant determined by the geometry of the cantilever. The parameter $k$ is the spring constant described by the geometry of the cantilever and the product materials. Thus, based on the (4.5), the relationship between the displacement $y$ and the actuated pressure difference $p$ can be easily derived.

The extended Bernoulli Equation (4.6) can be used to describe the static fluid flow in the gap between the cantilever and the valve seat

$$p_1 + \alpha\frac{\rho}{2}v_1^2 = p_2 + \alpha\frac{\rho}{2}v_2^2 + (p_e)_{1-2} \tag{4.6}$$

where $p_1$,$v_1$ and $p_2$, $v_2$ are the pressure and velocity at the beginning and the end of the gap. $\alpha$ is a kinetic energy coefficient relevant to the velocity profile. It is assumed to be 2.0 due to the laminar flow in a channel. The parameter $(p_e)_{1-2}$ describes the friction loss.

To derive an analytical result, the gap between the cantilever and the valve seat is divided into five pieces (Figure 4.16). In studying the functions of the pressure difference $p$ and the displacement of cantilever beam at the individual regions, the overall flow rate $\Phi$ can be treated as a function of pressure difference $p$ and the displacement $y$:

$$p = \sum_{i=I}^{V} \Delta p_i(\Phi, y) \tag{4.7}$$

Therefore [16],

$$\Phi(p, y) = \frac{\rho\alpha}{b^2 s^2}\left[-\frac{12\mu}{b}\left(\frac{l_1}{y^3} + \frac{l_1}{s^3}\right) + \sqrt{\frac{144\mu^2}{b^2}\left(\frac{l_1}{y^3} + \frac{l_2}{s^3}\right)^2 + \frac{2\rho}{b^2}\frac{\alpha p}{s^2}}\right] \tag{4.8}$$

By substituting $y = f(p)$ , $s = y + h$ in the above equation, it can be shown that the static flow rate is fully determined by the actuated pressure difference and the structural parameters, $\Phi = g(p, y_1, y_2, ..., y_n)$ [16]. The analytical model of the microvalve is studied with SystemC. Figure 4.17 shows the microvalve model coded by VHDL-AMS and SystemC, respectively. Because SystemC does not provide an associated simulator, the simultaneous ODAEs are combined with a ODAE solver coded by users. These functions are solved by a *Process: ODAEs()*.

Table 4.2 shows the microvalve determined design parameters and their design value. In addition, the fluid density and viscosity for each fluidic sample are assumed to be the same. By setting the microvalve operating frequency at $100Hz$, the average flow rate of the microvalve can be calculated to be 5.86 $ml$ per minute.

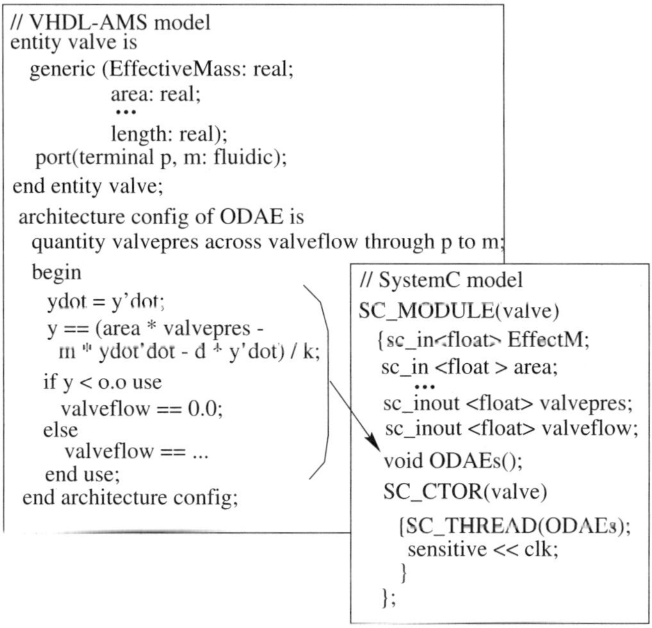

```
// VHDL-AMS model
entity valve is
  generic (EffectiveMass: real;
           area: real;
           ...
           length: real);
  port(terminal p, m: fluidic);
end entity valve;
architecture config of ODAE is
  quantity valvepres across valveflow through p to m;
  begin
    ydot = y'dot;
    y == (area * valvepres -
      m * ydot'dot - d * y'dot) / k;
    if y < o.o use
      valveflow == 0.0;
    else
      valveflow == ...
    end use;
end architecture config;
```

```
// SystemC model
SC_MODULE(valve)
  {sc_in<float> EffectM;
   sc_in <float > area;
   ...
   sc_inout <float> valvepres;
   sc_inout <float> valveflow;
   void ODAEs();
   SC_CTOR(valve)
     {SC_THREAD(ODAEs);
      sensitive << clk;
     }
  };
```

**FIGURE 4.17**

**Microvalve model coded in VHDL-AMS and SystemC. The simultaneous ODAEs in VHDL/AMS are combined with a ODAE solver, and are solved by a process *ODAEs()*.**

**Table 4.2** Elemental Parameters and Initial Nominal Design Values

| Parameters | Values | Units |
|---|---|---|
| Length of the cantilever ($L$) | 1600 | $\mu m$ |
| Width of the cantilever($b'$) | 1000 | $\mu m$ |
| Thickness of the cantilever ($h'$) | 15 | $\mu m$ |
| Height of the valve seat ($h$) | 50 | $\mu m$ |
| Length of the valve seat ($l_1$) | 5 | $\mu m$ |
| Width of the valve seat ($b$) | 400 | $\mu m$ |
| Length of the cantilever over valve seat($l_2$) | 100 | $\mu m$ |
| Young's Modulus ($E$) | 146.9 | GPa |
| Air Pressure($P_a$) | 100000 | $P_a$ |

## 4.4 System Performance Analysis and Design Optimization

We combine the system simulation with the stochastic macro model and component level simulation as discussed in Section 4.3. The hierarchical simulation results with SystemC are shown in Figures 4.18–4.20. These results show the system performance when there are 100 fluidic samples entering the system for processing. The performance analysis metrics include throughput, resource utilization, and execution time distributions. These data are useful in identifying how component performance metrics impact overall architectural performance, and they provide guidance for optimization.

Figure 4.18 shows the system throughput for three kinds of fluidic samples. Figure 4.19 shows the system resource utilization for processor elements: analyzers, mixers, and catalyzers, as well as the microchannel. The total time for each fluidic sample staying in the handling system consists of three periods: the time of waiting for system resources, the processing time, and the microchannel delivering time. Figure 4.20 shows the average values over three periods for each kind of fluidic sample.

### 4.4.1 Architectural Optimization

Table 4.3 shows storage utilization statistics for the containment reservoirs and the bus storage buffer. Partitioning storage between "on/off chip" allows tradeoffs in area, speed, and costs. The maximum number of occupied cells in channel bus storage buffer is equal to $N_{pr} + 1$, where $N_{pr}$ is the total number of processors, and it is set to 9. The average number of occupied cells is 9.1584; this shows that the

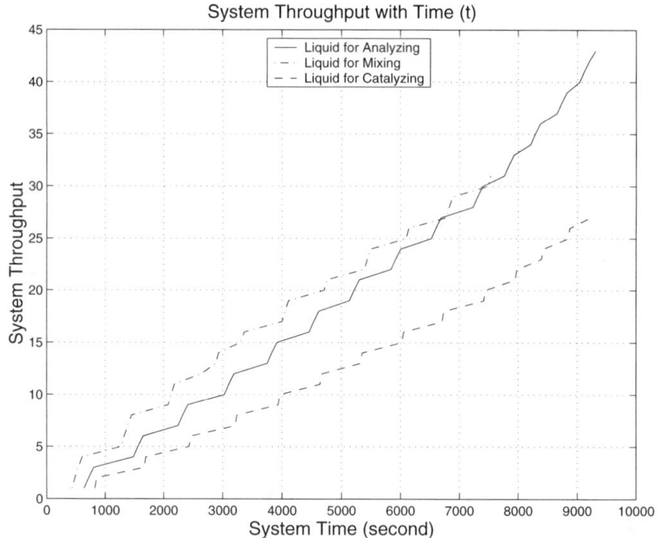


**FIGURE 4.18**
**Micro-chemical handling system throughput for three kinds of fluidic samples.**

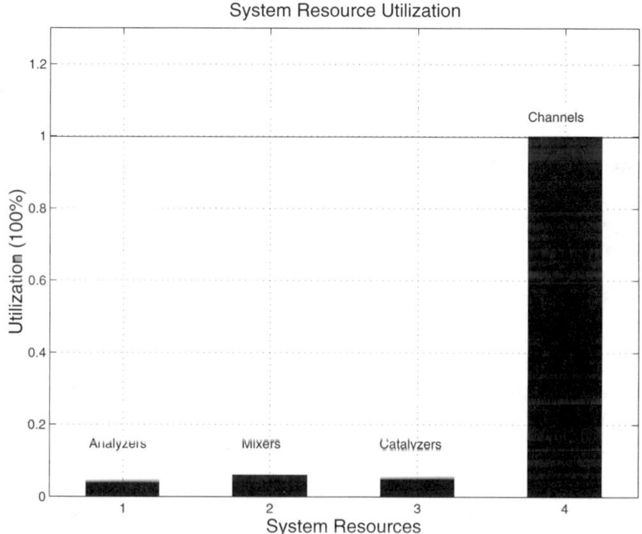


**FIGURE 4.19**
**Micro-chemical handling system resource utilization for three processors and a microchannel.**

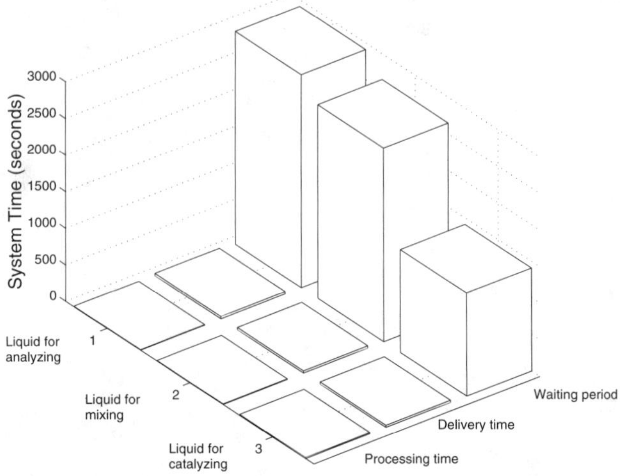


**FIGURE 4.20**
**Average system time distribution for each kind of fluidic sample. The system time includes: the time of waiting for system resources, the processing time, and the microchannel delivering time.**

scheduling scheme achieves good utilization.

**Table 4.3** Microsystem Simulation Summary

| | Analyzing Reservoir | Mixing Reservoir | Catalyzing Reservoir | Storage Buffer |
|---|---|---|---|---|
| Average Number of Occupied Cells | 13.2326 | 7.9032 | 8.8889 | **9.1584** |
| Maximum Number of Occupied Cells | 30 | 19 | 18 | **10** |

The arriving fluidic sample is stored in an appropriate containment reservoir and it is processed by the associated processor depending on its purpose. Figure 4.19 shows that the channel bus is fully utilized, and processor utilization shows resource availability. Figure 4.20 shows that the chemical fluidic sample waiting time occupies a large proportion of the total cycle time. The channel bus is the principal bottleneck preventing liquid samples from accessing processors and processed liquid samples from being dispensed. The effects of increasing channel bus bandwidth by adding another parallel channel bus are studied in Figures 4.21 and 4.22. Figure 4.21 compares system throughput versus channel bus bandwidth, and Figure 4.22 compares system resource utilization versus channel bus bandwidth. The improved architecture with the two-channel-bus architecture reduces system processing time and increases resource availability.

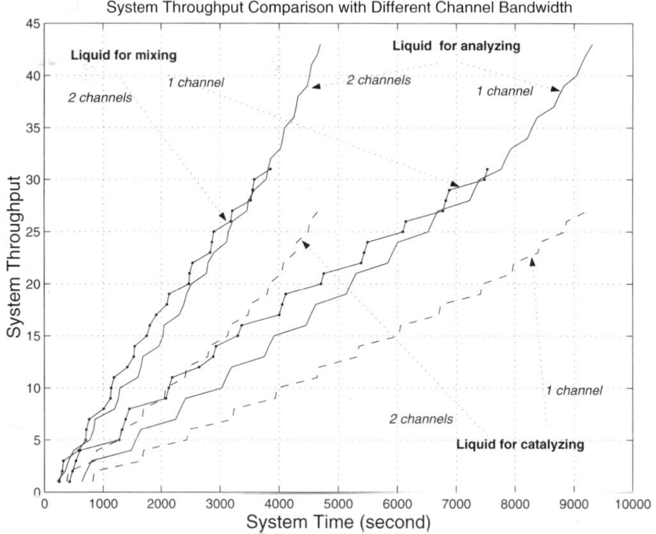


**FIGURE 4.21**
**System throughput for each kind of fluidic sample versus channel bus bandwidth. Two-channel-bus architecture reduces the system processing time.**

## 4.4.2 Microsystem Performance Sensitivity Analysis

The variation of the microvalve actuation frequency can change the flow rate, which, in turn, can affect microsystem performance. Typically miniature biomedical and chemical applications involving small liquid volumes require high and stable processing throughput. Thus, the micropump flow rate, based on the microvalve actuation frequency, is an important system-level design parameter influencing performance.

Considering the operating frequency tolerance, two frequencies are selected with forward and backward tolerance $10Hz$, respectively, assumed to be the worst-case micropump actuation frequency $(f)$ variation. Table 4.4 shows the results of analyzing micro-chemical handling system performance sensitivity with three different frequency combinations.

The above simulation results show that micropump flow rate is tightly correlated to the micropump operating frequency, which, in turn, impacts the system performance

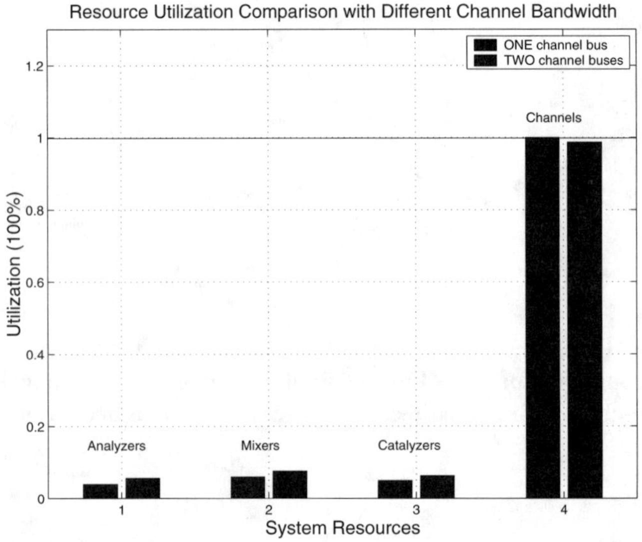

Resource Utilization Comparison with Different Channel Bandwidth

ONE channel bus
TWO channel buses

Utilization (100%)
System Resources

Analyzers
Mixers
Catalyzers
Channels

**FIGURE 4.22**
**System source utilization versus channel bus bandwidth. The left side bar indicates the utilization for the one-channel architecture, and the right side bar presents the utilization for the two-channel architecture. Two-channel-bus architecture increases resource availability.**

**Table 4.4** System Performance Comparison due to Different Operating Frequency

| Statistical Parameters | Actuation Frequency Selection | | |
|---|---|---|---|
| | $f = 90Hz$ | $f = 100Hz$ | $f = 110Hz$ |
| Total System Time(t) Required for 100 throughputs | **5144** | **4690** | **4282** |
| Analyzers Utilization (%) | 5.7 | 5.9 | 6.2 |
| Mixers Utilization(%) | 8.1 | 7.6 | 7.8 |
| Catalyzers Utilization (%) | 5.7 | 6.3 | 6.9 |
| Channel Utilization (%) | 98.3 | 98.7 | 98.2 |

and throughput.

### 4.4.3 Microsystem Performance Estimation with Traffic Variation

Acquisition rate (workload) is another important system-level design parameter influencing system performance. For a given architecture, the micro-chemical handling system has a saturation capacity, where resources are maximally utilized. Workloads less than saturation capacity under-utilize resources, whereas workloads greater than saturation capacity may decrease system quality or even cause system failure. Thus, it is desirable to investigate saturation operating performance.

As explained in Section 4.4, liquid sample acquisition rate is modeled by an exponential probabilistic distribution, governed by (4.1). Nominal mean interarrival time is assumed to be 15 seconds, i.e. $\lambda = \frac{1}{15}$. Figures 4.23 shows the micro-chemical handling system performance under varying sample acquisition rates, $\lambda$ changing from $\frac{1}{100}$ to $\frac{1}{5}$.

Figure 4.23 shows the system processing capability versus different traffic rates. The vertical axis presents the system processing capability, denoted by using the number of processed fluidic samples per hour. Here the basic time unit is a second. The horizontal axis represents the sample traffic rate, $\lambda$, meaning the number of fluidic samples arriving per second. The straight dash-point line shows an ideal case when availability of the system is guaranteed and all input samples are accepted and processed. The other curve is actual system performance. When the sample traffic rate is low, system throughput is nearly linear – the performance of the system approximates the ideal. At increased input rates, meaning reduced interarrival time, actual system processing capability increases and quickly reaches saturation. Figure 4.24 shows that increasing the acquisition rate (increasing $\lambda$), but not beyond system saturation capacity, improves performance by increasing system resource utilization. At saturation, throughput remains constant regardless of input rate variation, but, as shown in Figure 4.25, queuing demands increased the requirement for storage buffer

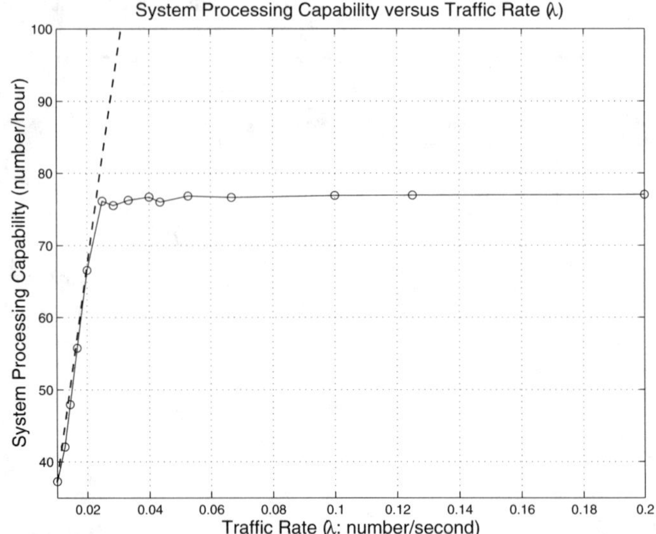


**FIGURE 4.23**
**System processing capability versus different traffic rate $\lambda$. After the system reaches saturation, system processing capability (throughput) remains constant regardless of input rate variation.**

capacity.

## 4.5 Conclusion

In this chapter, we have presented a hierarchical modeling and simulation methodology that combines high-level stochastic queuing networks with low-level nodal conservative differential equations. The complete system modeling and simulation for a micro-chemical handling system is presented depending on the SystemC integrated design environment. Simulation results and performance analysis data have been presented. Performance analyses investigate throughput, resource utilization, and system architectural optimization. The influence of micropump actuation frequency and input liquid sample acquisition rate have also been studied. The results of this work assist in understanding the complexities of system performance and serve to guide next-generation designs. The detailed circuit-level MEFS hierarchical design optimization/verification methodology is discussed in Chapter 5.

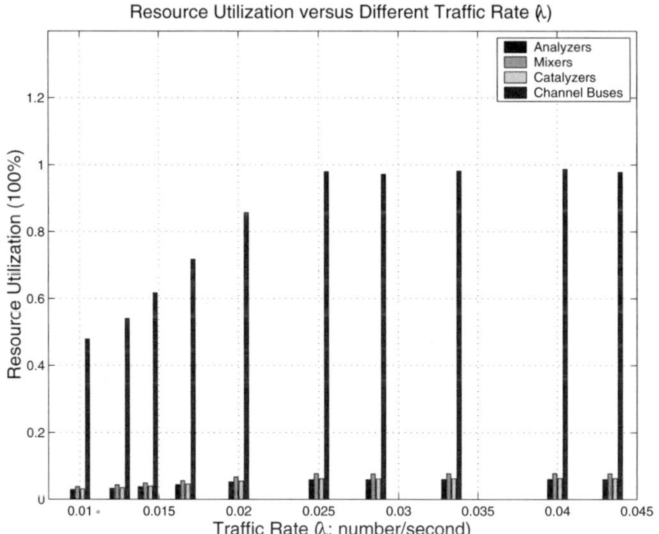


**FIGURE 4.24**

**Resource utilization versus different traffic rate $\lambda$. Each bunch of bars indicates the utilization of system resources at certain traffic rate. Each bar for every group, from left to right, represents the utilization of analyzers, mixers, catalyzers, and the channel bus, respectively.**

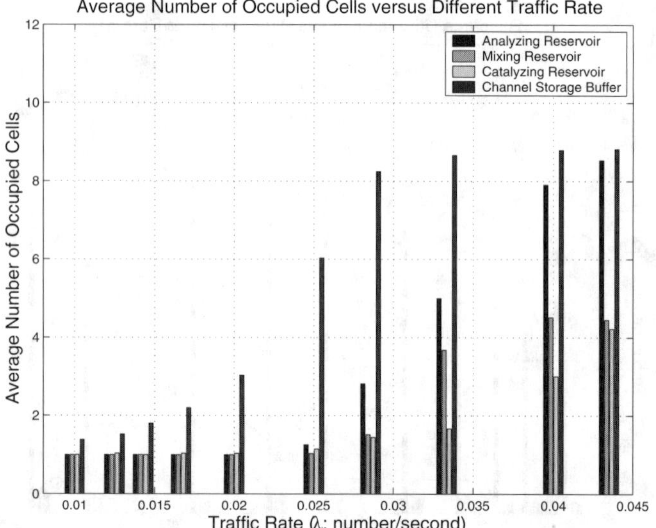


**FIGURE 4.25**

**Average number of occupied containment cells versus different traffic rate $\lambda$. Each bunch of bars indicates the number of occupied cells in containment chambers at certain traffic rate. Each bar for every group, from left to right, represents the analyzing reservoir, mixing reservoir, catalyzing reservoir, and the channel storage buffer, respectively.**

# Chapter 5

# Circuit-level Optimization

A hierarchical integrated design optimization approach for MEFS includes five main components. The content is illustrated in Figure 5.1 and the components are listed below.

1. Simulation design methodology
2. Optimization verification
3. On-target design methodology
4. Robust design methodology
5. Application flexibility optimization methodology

In the following sections, each of these components is described in detail. In Section 5.1, two efficient simulation methods are studied: the bootstrap method and the factorial design method. They are used in our optimization approaches. Section 5.2 discusses verification methods that verify the correctness of optimal design results. From Section 5.3 to 5.5, three optimization methodologies are demonstrated for microsystem design and process optimization. A statistical response analysis strategy is proposed in Section 5.3. This strategy can efficiently find an on-target design point that meets the performance goals. It also benefits to the optimal control for fabrication and operation. In Section 5.4, by studying the relationships between the design parameters and system performance, a robust design methodology is demonstrated based on the Taguchi experiment method. Moreover, in order to overcome the limitation of "custom microsystems", and to extend the range of MEFS performance, a reconfigurable microsystem design methodology is demonstrated along the lines of hardware/software co-design to achieve functional unit reusability. This design methodology for application flexibility is presented in Section 5.5.

## 5.1 Simulation Design Methodology

Due to the large number of design parameters and coupled-energy interactions, time-consuming numerical methods for DAEs are necessary to study MEFS system behavior. Simulation cost is becoming one of the main obstacles for system designers. Although traditional Monte Carlo simulation methods have been widely used in electronic design [78], it is infeasible for microsystem optimization due to the expensive computational cost. The bootstrap method [79] and the factorial design method [31] are more efficient for composite microsystems.

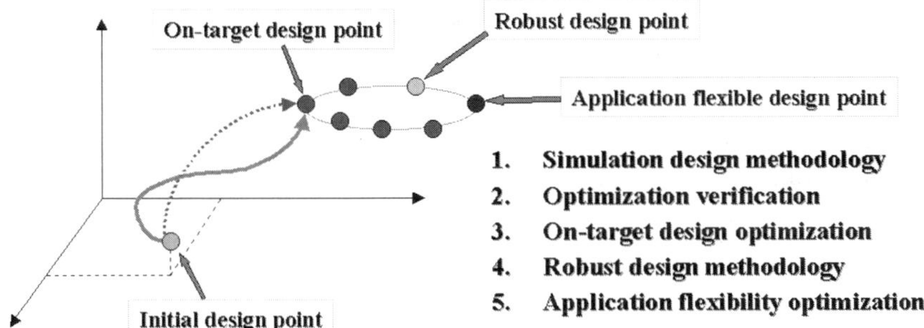


**FIGURE 5.1**
**Hierarchical integrated design optimization includes five components: Simulation design methodology, Optimization verification, On-target design optimization, Robust design methodology, and Application flexibility optimization.**

### 5.1.1 Bootstrap Method

Some simulation techniques compute either the variance of parameter estimators or confidence intervals for the true parameters. This approach requires a sufficiently large sample size so that the "asymptotic" result can be applied to study the system behavior accurately. However, large sample size increases computational complexity. As an alternative to Monte Carlo simulation, the bootstrap method can be used with a smaller sample space.

The bootstrap method is a well-established robust statistical methodology [79]. It can assess the accuracy of a parameter estimator in situations where conventional techniques are not feasible, and it is potentially superior to large-sample techniques. With 10 simulation results, the bootstrap method can get the same accuracy with around 100 runs of Monte Carlo simulation [80]. The bootstrap has already been widely used in signal processing [81]. Composite microsystems is its new application area.

The basic principle for the bootstrap method is the random replication of the original sample. In order to obtain a parameter estimator $S$ using a sample $\mathbf{x}$, where $\mathbf{x} = \{x_1, x_2, ..., x_n\}$, the bootstrap method includes the following steps.

- Conduct the experiment to obtain the random sample $\mathbf{x} = \{x_1, x_2, ..., x_n\}$. The $x_i$'s are independent and identically distributed random variables.
- Generate a bootstrap sample, called the bootstrap resample

  $$\mathbf{X^B} = \{x^1, x^2, ..., x^m\}$$

  where

  $$\mathbf{x^b} = \{x_1^b, x_2^b, ..., x_n^b\}, \quad b \in \{1, 2, ..., m\}$$

$\mathbf{x^b}$ has the same dimension as $\mathbf{x}$, and the elements of $\mathbf{x^b}$ are randomly drawn from the original sample with replacement. For example, if we have a set $\mathbf{x} = \{1.0, 3.0, 4.3, 3.4\}$, one of the bootstrap sample might be $\mathbf{x^b} = \{1.0, 1.0, 4.3, 3.4\}$.

- Approximate the distribution of the mean and variance of $\mathbf{x}$ by the distribution of the mean and variance derived from $\mathbf{X^B}$. A series of new statistical estimates can be calculated from $\mathbf{X^B}$ as follows:

$$S^B = \{S^1, S^2, ..., S^b, ..., S^m\}$$

and the robust estimate of $S$ is

$$S^* = E(S^B).$$

In the following section, the application of the bootstrap method is illustrated with a statistical parametric regression analysis of a special composite microsystem: microelectromechanical resonator.

#### 5.1.1.1 Parametric Analysis With Bootstrap Method

With the number of applications of integrated composite microsystems growing, there is a need to address issues of the performance, manufacturing yield, and operational reliability. These issues require detailed and systematic analyses of composite microsystems. These analyses characterize the relationships between basic design parameters and overall design function and performance. Due to the complexity of composite microsystems in terms of the number of components and the coupled-energy interactions, these analyses are investigated using a statistical parametric regression analysis approach.

Microelectromechanical resonators are selected as representative of composite microsystems due to their broad and flexible use. In this special case study, the bootstrap method is used to analyze complex systems by conducting sampled simulations. The simulation result is compared with traditional Monte Carlo simulation methods. In addition, various sets of input parameters and corresponding system states are perturbed using deterministic or stochastic techniques. The resulting system behavior forms a sampled envelope describing overall function and performance. For microelectromechanical resonators, the effects of perturbing several electrical and mechanical parameters on circuit operation are studied. The bootstrap method reveals first order sensitivities and correlations that are useful in design optimization.

The following sections explain the physical principles governing a linear comb drive microresonator and present the simulation model, based on equivalent circuit microelectromechanical systems modeling. Sample circuit simulation results are reported. The concept of parametric regression analysis is explained and extensive bootstrap method results are presented that characterize the influence of variations of basic mechanical/electrical parameters on nominal microresonator performance. The estimation of the variance of the parameter estimate of the regression process are compared. Finally, findings of the parametric regression analysis are given.

#### 5.1.1.2 System Modeling of Comb-drive Microresonator

Early work on microresonators and the folded-flexure electrostatic comb drive microelectromechanical resonator, as shown in Figure 5.2, was conducted by Howe and Tang [82]. The device consists of a movable central shuttle mass that is suspended above the substrate by folded flexure beams. The folded flexure beams are anchored on the substrate at two central points, allowing the shuttle mass to move laterally - parallel to the substrate surface. Transverse motion is also possible, but is restrained by the folded flexure beam and ground planes. The folded flexure beams and shuttle mass can be fabricated using basic integrated circuit lithographic processes involved in surface micromachining. A sacrificial or spacing layer removed during fabrication provides for suspension off the substrate surface for deflectional motion.

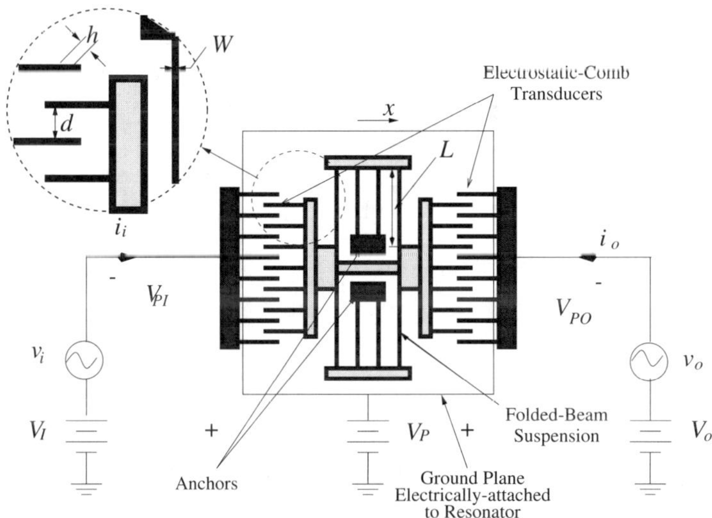


**FIGURE 5.2**
**Two-port, folded-beam, lateral comb-driven resonator consists of a movable central shuttle mass, two folded flexure beams.**

Linear comb drives are formed by interdigitated fingers placed on each half of the shuttle mass. The resulting structure uses the relationship between electrostatic force and capacitance to create sustained oscillation. Exciting a linear comb capacitor by a time varying voltage $V(t)$ produces a time varying electrostatic force $f(t)$ [83]

$$f(t) = \frac{\partial U(V(t), x)}{\partial x} \tag{5.1}$$

where $U$ denotes electric field energy. The time varying electrostatic force sets the shuttle mass into vibratory motion, with the strongest fundamental mode setting the resonating frequency. This motion excites another linear comb capacitor to produce a time varying capacitance ($C$), which, in turn, yields a time varying current ($i(t)$).

$$i(t) = C(t)\frac{\partial V(t)}{\partial t} + V(t)\frac{\partial C(t)}{\partial t} \tag{5.2}$$

The time varying current is fed back in phase to produce the original time varying voltage and closed loop oscillation is sustained.

The simulation model for the linear comb drive micromechanical resonator is given in Figures 5.3 and 5.4. This model is based on the work reported by Howe [84, 85], and built by using the equivalent circuit behavioral modeling technique.

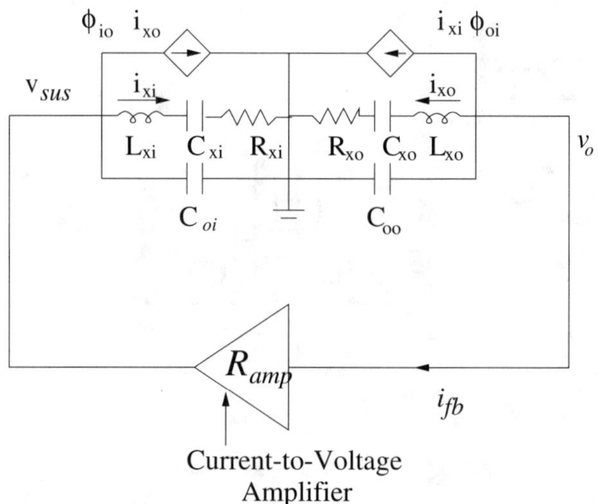


**FIGURE 5.3**
**Equivalent circuit for a two-port $\mu$resonator**

The basic mechanical resonating property is given by the second-order actuation force to resultant displacement transfer function shown in phasor notation

$$\frac{X(\jmath\omega)}{F(\jmath\omega)} = \frac{k_{sys}^{-1}}{1 - \left(\frac{\omega}{\omega_0}\right)^2 + \jmath\left(\frac{\omega}{Q\,\omega_0}\right)} \tag{5.3}$$

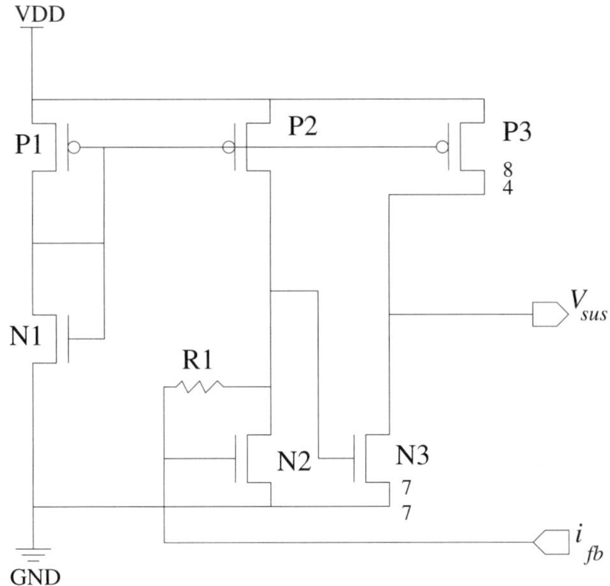


**FIGURE 5.4**
**Circuit schematic for sustaining amplifier, presenting the functionality of $R_{amp}$.**

where $X$ is the displacement, $F$ is the force, $k_{sys}$ is the effective system spring constant, $\omega$ is the radian frequency, $\omega_o$ is the resonance frequency, and $Q$ is the quality factor.

The displacement/force transfer function can be used to derive the transfer function relating input drive voltage to resulting motional current $I_o/V_i$ by using the expressions given in (5.4) and (5.5) and neglecting the effects of DC and higher-order frequency components.

$$\frac{F(\jmath\omega)}{V_i(\jmath\omega)} \approx -V_{PI}\,\frac{\partial C}{\partial x} \tag{5.4}$$

$$\frac{I_o(\jmath\omega)}{X(\jmath\omega)} \approx V_{PO}\,\jmath\omega\,\frac{\partial C}{\partial x} \tag{5.5}$$

The resulting transfer function yields the values for the series resonant circuit composed of $L_x$, $C_x$, and $R_x$, given respectively by (5.6)-(5.8) [83].

$$L_x = \frac{k_{sys}}{\omega_o^2 V_{PI}^2 (\frac{\partial C}{\partial x})^2} \tag{5.6}$$

$$C_x = \frac{V_{PI}^2}{k_{sys}} (\frac{\partial C}{\partial x})^2 \tag{5.7}$$

$$R_x = \frac{k_{sys}}{\omega_o Q V_{PI}^2 (\frac{\partial C}{\partial x})^2} \tag{5.8}$$

The values of the $L, C, R$ series resonant circuit elements are determined by resonator geometry and bias voltage.

Expressions for the fundamental lateral resonance frequency ($\omega_o$) and quality factor ($Q$) are given in (5.9) and (5.10) [86].

$$\omega_o = \left[\frac{2Eh(W/L)^3}{(M_p + 0.3714M)}\right]^{\frac{1}{2}} \tag{5.9}$$

$$Q = \frac{z}{\mu A_p}(M_p k_{sys})^{\frac{1}{2}} \tag{5.10}$$

Where $E$ is the Young's modulus, $M_p$ is the mass of the shuttle plate, $A_p$ is the area of the shuttle plate, $M$ is the mass of the supporting beams, $\mu$ is the absolute viscosity of air, and $z$ is the separation distance between folded-flexure linear comb drive resonator and the substrate. The dimensional definitions for $h$, $W$, and $L$ are illustrated in Figure 5.2.

Neglecting fringing fields, the total capacitance of the drive-side linear comb structure denoted by $C_1$ is given by

$$C_1(x) = \frac{2N_1\varepsilon_0 h(L_1 + x)}{d_1} \tag{5.11}$$

where $N_1$ is the number of comb drive fingers on the drive side (port 1), $\varepsilon_0$ is the permittivity of free space, $h$ is the finger thickness, $L_1$ is the finger overlap, and $d_1$ is the finger gap.

Combining (5.6)-(5.8) and 5.11, and realizing the symmetry of the linear comb drive microresonator yields the expressions for $L_{xi}$, $L_{xo}$, $C_{xi}$, $C_{xo}$, $R_{xi}$, and $R_{xo}$ given in the following:

$$C_{oi} = \frac{2N_1\varepsilon_0 hL_1}{d_1} \qquad C_{oo} = \frac{2N_2\varepsilon_0 hL_2}{d_2}$$

$$L_{xi} = \frac{k_{sys}}{4}\left[\frac{d_1}{\omega_r N_1\varepsilon_0 hV_{P1}}\right]^2 \qquad L_{xo} = \frac{k_{sys}}{4}\left[\frac{d_2}{\omega_r N_2\varepsilon_0 hV_{P2}}\right]^2$$

$$C_{xi} = 4k_{sys}^{-1}\left[\frac{N_1\varepsilon_0 hV_{P1}}{d_1}\right]^2 \qquad C_{xo} = 4k_{sys}^{-1}\left[\frac{N_2\varepsilon_0 hV_{P2}}{d_2}\right]^2$$

$$R_{xi} = \frac{k_{sys}}{4\omega_r Q}\left[\frac{d_1}{N_1\varepsilon_0 hV_{P1}}\right]^2 \qquad R_{xo} = \frac{k_{sys}}{4\omega_r Q}\left[\frac{d_2}{N_2\varepsilon_0 hV_{P2}}\right]^2$$

$$\phi_{oi} = \frac{d_1 N_2 V_{PO}}{d_2 N_1 V_{PI}} \qquad \phi_{io} = \frac{d_2 N_1 V_{PI}}{d_1 N_2 V_{PO}}$$

The capacitors $C_{oi}$ and $C_{oo}$ denote conventional feeding through current due to the DC capacitance of the linear comb drives. The current amplification factors $\phi_{oi}$ and $\phi_{io}$ account for possible drive/sense asymmetries in bias voltages, number of digits, and inter-digit gaps. Figures 5.5-5.7 show the simulation results of the microresonator.

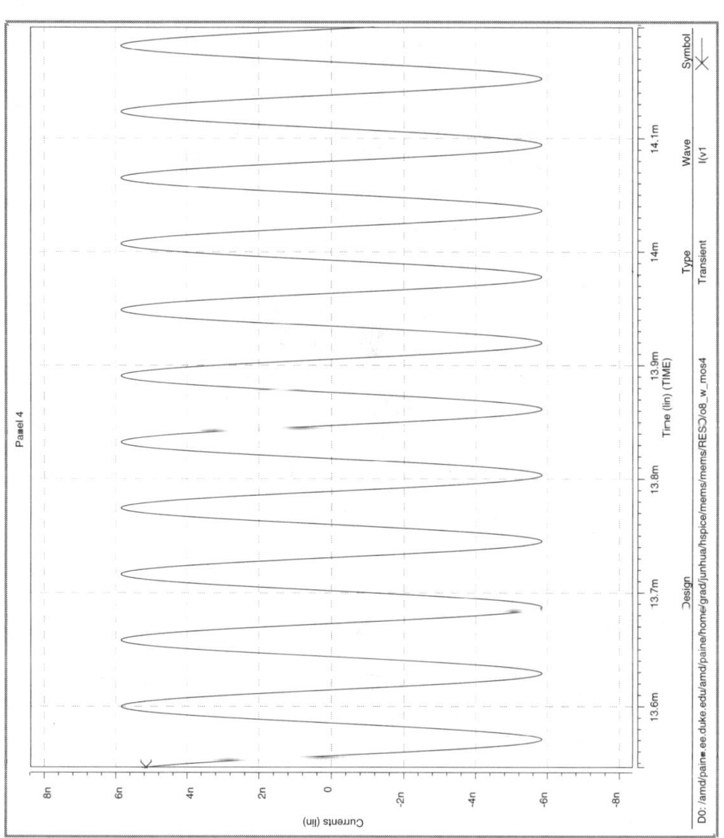


**FIGURE 5.5**
**Steady state microresonator oscillation**

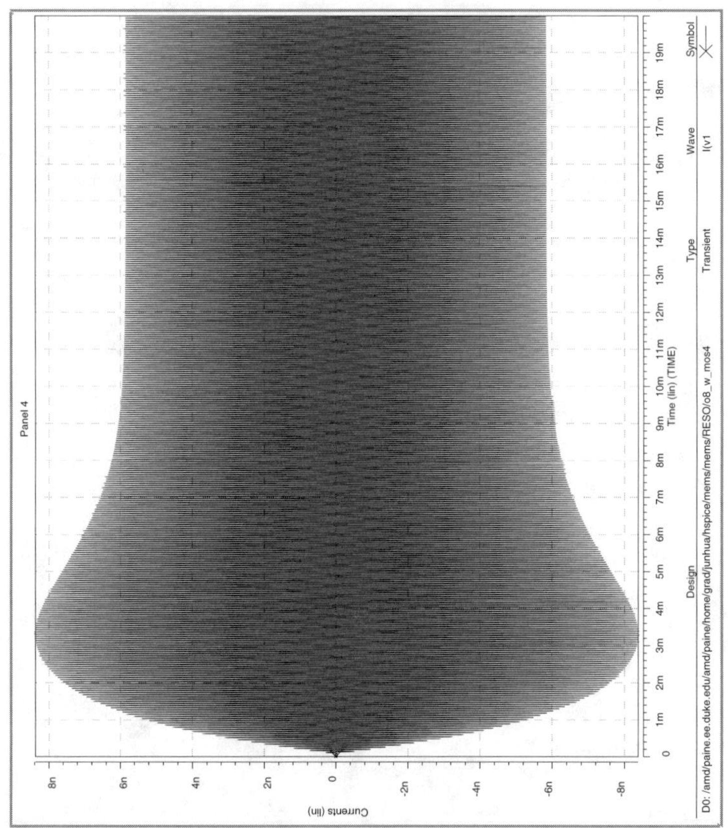


**FIGURE 5.6**
**Transient microresonator oscillation**

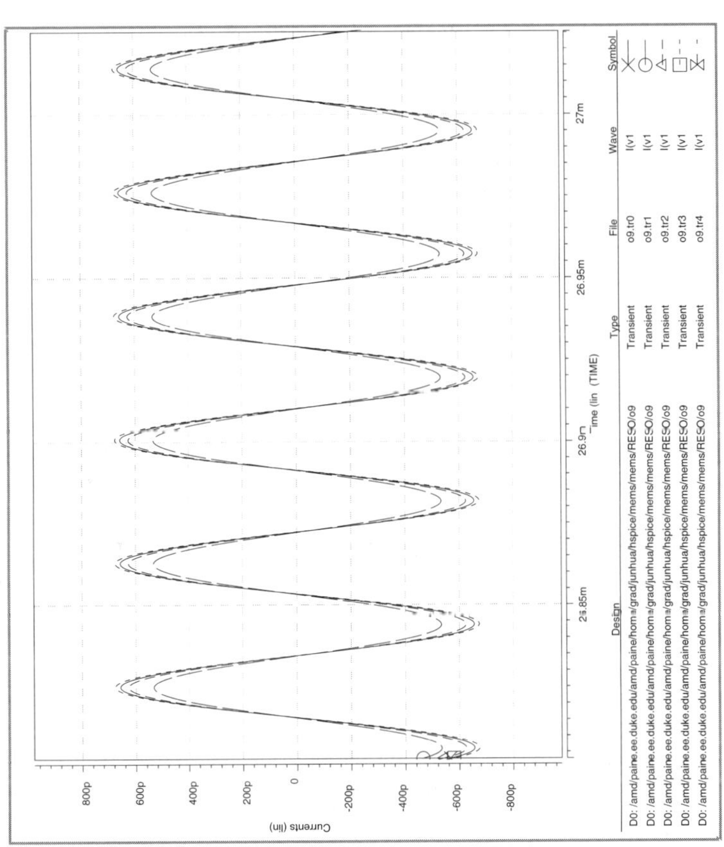

**FIGURE 5.7**
**Waveform for current due to the variance of $R_1$**

#### 5.1.1.3 Simulation and Statistical Regression Analysis

The circuit model of the microelectromechanical resonator is used in the bootstrap method to study the effects of basic parametric variations on microresonator operation. The basic design parameters and nominal values are listed in Table 5.1 [87].

**Table 5.1** Elemental Parameters and Nominal Design

| Parameters | Values | Units |
|---|---|---|
| Width ($W$) | 2 | $\mu m$ |
| Length($L$) | 200 | $\mu m$ |
| Thickness of the fingers ($h$) | 2 | $\mu m$ |
| Gap ($d$) | 2 | $\mu m$ |
| Number of the fingers ($N$) | 12 | |
| Young's Modulus ($E$) | 150 | GPa |
| Bias Voltage ($V_P$) | 80 | V |

Microresonator operation is measured by the design metrics of resonant frequency $\omega_0$ and linear comb drive motional transconductance gain $Y_x$. The resonant frequency is an important design performance. The linear comb drive transconductance gain is also a critical design objective for operation in that, as a measure of the generated sense current per a given drive voltage, motional transconduction gain indicates the effectiveness of the overall transduction. A higher transconductance gain reflects a better linear comb drive resonator.

The bootstrap method uses the statistical analysis to determine sensitivity gradients. Sensitivity gradients give the degree of dependence of the system's expected output on input parameters. There are several gradient-estimation methods; the following statistical analyses employ linear regression [88].

Linear regression conducts sample simulations, varying a parameter value over a given range. Then, the derivative is estimated by noting the change in the performance measure owing to the change in the corresponding parameter. Provided the variation ranges of the electrical/mechanical parameters are small, the statistical relationship between the input parameters and output performances can be considered linear and thus, described with a regression line model. The validity of linear regression in representing the actual relationship between two variables can be checked by computing the correlation. Correlation is a measure of the strength of the linear relationship between two variables.

The parametric regression analysis is conducted in three stages. First, mechanical and electrical parameters of the microelectromechanical resonator are systematically modified using a normal distribution and the effect on performance is examined using the bootstrap method. Then, linear regression is computed by least-squares curve fitting to obtain sensitivity gradient estimates. Finally, the correlation of the data is computed to check the validity of the data analysis.

Making the constraints of single parameter variations and statistical independence of parameter variations, the relationship between the input elemental parameters and output performance metrics, such as the influence of the single parameter $W$ on resonant frequency $\omega_o$, is given by

$$\omega_o = f_{11}(W) + K \tag{5.12}$$

where $f_{11}(W)$ denotes the influence of $W$ on $\omega_o$ and $K$ represents the influence of the other three parameters on resonant frequency. Given small variations of elemental parameters, (5.12) becomes a linear function

$$\bar{\omega_o} = b_0 + b_1 W \tag{5.13}$$

and the coefficients can be estimated using the theory of linear regression. The regression coefficients $b_0$ and $b_1$ indicate the relationship between the parameter $W$ and the performance $\omega_o$, with the slope $b_1$ indicating the degree of dependence.

Denoting the $(\omega_o, W)$ observations from the statistical simulation as $(\omega_{o1}, W_1)$ for the first simulation, $(\omega_{o2}, W_2)$ for the second trial, and $(\omega_{oi}, W_i)$ for the $i^{th}$ trail (where $i = 1, 2, \ldots, m$), (5.14) and (5.15) respectively give the computation for $b_1$ and $b_0$ using the least-squares method [89].

$$b_1 = \frac{\sum_{i=1}^{m}(W_i - \bar{W})(\omega_{ri} - \bar{\omega_o})}{\sum_{i=1}^{m}(W_i - \bar{W})^2} \tag{5.14}$$

$$b_0 = \frac{1}{m}\left(\sum_{i=1}^{m}\omega_{ri} - b_1 \times \sum_{i=1}^{m} W_i\right) \tag{5.15}$$

Since the linear regression method *approximately* represents the relationship between two variables with a straight line, it is necessary to check the validity of this linear approximation using the correlation coefficient. (5.16) and (5.17) give the formulas for calculating covariance and correlation, respectively [90].

$$Cov(X,Y) = \frac{\sum_{i=1}^{n}(x_i - \bar{x})(y_i - \bar{y})}{n} \tag{5.16}$$

$$Cor(X,Y) = \frac{Cov(X,Y)}{\sqrt{\frac{\sum_{i=1}^{n}(x_i - \bar{x})^2}{n} \frac{\sum_{i=1}^{n}(y_i - \bar{x})^2}{n}}} \tag{5.17}$$

#### 5.1.1.4 Output Analysis Report

Monte Carlo simulation analysis needs over 30 simulation results [87], while based on the bootstrap method principle, six simulation results with 1000 resamplings are used. Figures 5.8-5.11 show the simulation result of the bootstrap method comparing to that of Monte Carlo in [87]. This comparison shows that the bootstrap method can reduce the computational cost while providing good parameter estimates.

Table 5.2 presents the linear regression analysis and associated correlations for microresonator resonant frequency $\omega_o$. Related figures are shown in Figures 5.8-5.11.

**Table 5.2** Linear Regression Analyses for $\omega_o$

| Parameters | Linear Function | Correlation |
|---|---|---|
| Width ($W$) | $\omega_o$=-5.446424+ 12.0099$W$ | 0.9936 |
| Length ($L$) | $\omega_o$=52.436760- 0.1605$L$ | 0.9024 |
| Number of the fingers ($N$) | $\omega_o$=22.834774 - 0.27985$N$ | 0.8704 |
| Young's Modulus ($E$) | $\omega_o$=9.814833 + 0.05506$E$ | 0.999 |

Table 5.2 shows that the correlation for each parameter with the performance metric $\omega_o$ is in an acceptable range, except $N$. Hence, the linear regressions appropriately express the relationships between the input parameters and output performances, and first-order sensitivities can thus be derived.

Tables 5.3 presents ranking of first-order parametric sensitivity gradients for the design metrics of resonant frequency. Most sensitive design parameters must be carefully controlled during design and fabrication. This information is very helpful to develop CAD tools for yield optimization of complicated composite microsystems.

**Table 5.3** Parametric Sensitivity Gradients for $\omega_o$

| Parameters | Resonant Frequency ($\omega_o$) |
|---|---|
| Width ($W$) | 1 |
| Length ($L$) | 2 |
| Number of the fingers ($N$) | – |
| Young's Modulus ($E$) | 3 |

The bootstrap method, as an alternative to Monte Carlo simulation, can be used to an-

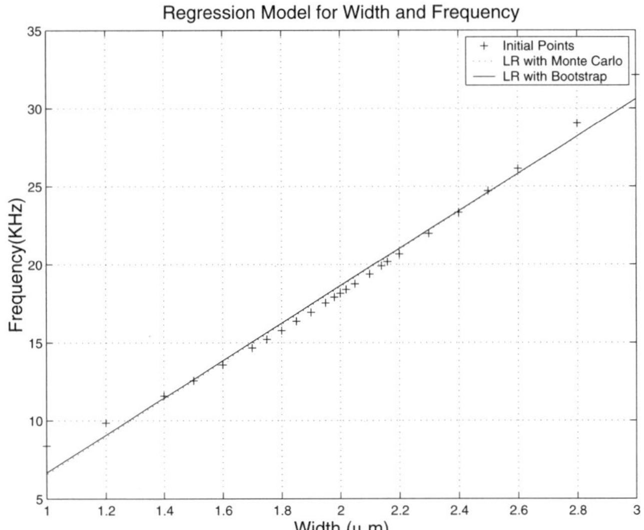


**FIGURE 5.8**
**Linear regression model for $W$ to $\omega_o$. The Bootstrap model matches the Monte Carlo model.**

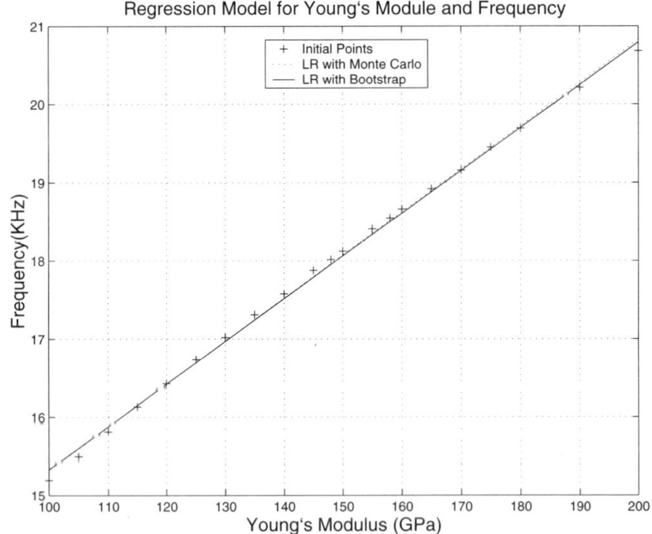


**FIGURE 5.9**
**Linear regression model for $E$ to $\omega_o$. The Bootstrap model matches the Monte Carlo model.**

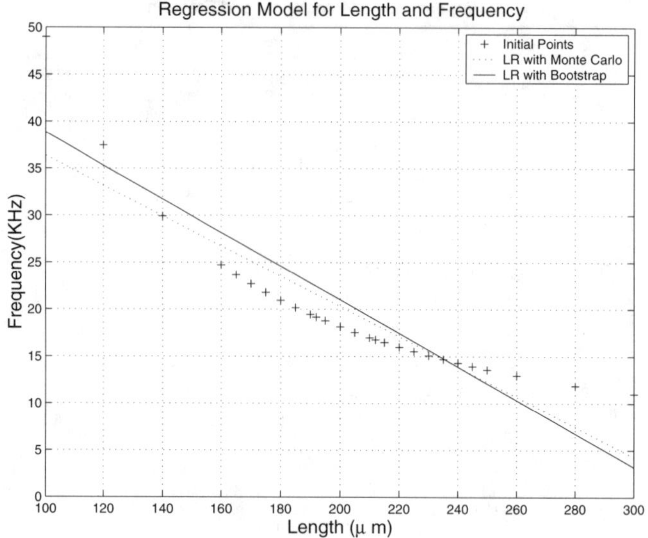


**FIGURE 5.10**
**Linear regression model for $L$ to $\omega_o$ shows the comparison between the results from the Monte Carlo method and that from the Bootstrap method.**

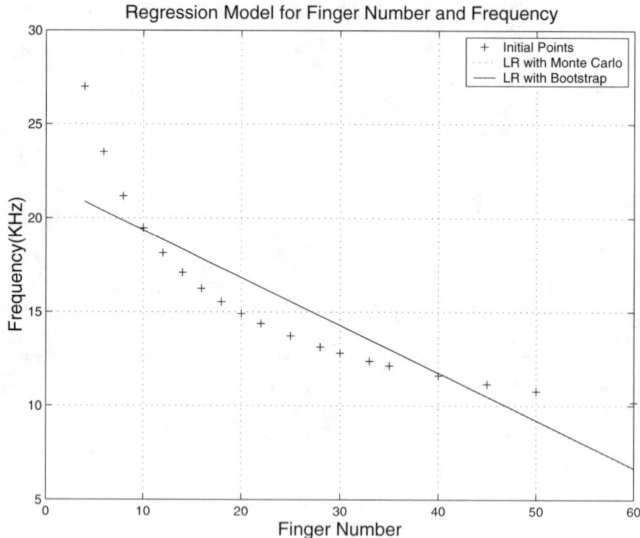


**FIGURE 5.11**
**Linear regression model for $N$ to $\omega_o$. The Bootstrap model matches the Monte Carlo model.**

alyze the sensitivity gradient of key electrical/mechanical characteristics on system performance. This first-order information about the trend of performance changes due to the variance of input parameters is useful for engineering design and the manufacturing process.

### 5.1.2 Factorial Design

Factorial design methods [91], which provide efficient techniques for systematically conducting and categorizing experiments, are always used to study the influence of multiple factors (design parameters) on the system performance. Factorial design methods include the complete factorial design method and the fractional factorial design method.

#### 5.1.2.1 Complete Factorial Design Methodology

The $2^k$ (or $3^k$) factorial design method limits each of the $k$ factors to take on only two (or three) levels: high and low (and nominal). By convention, $+1$ indicates the high level, $-1$ indicates the low level, and $0$ indicates the normal design point for the three-level factorial design. Level values are specified based on the design region, and they reflect the variance of the associated design parameters. Each possible factor-level combination defines a design point, and the collective design points define a matrix simulations of the desired performance envelope.

After the factor levels have been set, matrix simulations are run at each design point to determine the effect of various parameters. Matrix simulations use special orthogonal arrays [92], which allow the effects of several parameters to be evaluated efficiently. Table 5.4 shows a complete factorial design matrix for $k = 2$.

**Table 5.4** Design Matrix for Two Design Parameters

| No. | $x_1$ | $x_2$ |
|---|---|---|
| 1 | -1 | -1 |
| 2 | -1 | 1 |
| 3 | 1 | -1 |
| 4 | 1 | 1 |

The number of system simulations, $n$, is equal to $2^k$ with $k$ design parameters, and these simulations are orthogonal to each other. Hence, the complete factorial design methodology is often infeasible due to $O(2^k)$ computational complexity. The number of experiments increase exponentially with an increase in the number of design parameters. This motivates the fractional factorial design method.

#### 5.1.2.2 Fractional Factorial Design Methodology

The number of simulations associated with factorial design can be reduced by using fractional factorial design methods. Fractional factorial design methods use orthogonal arrays to screen experiments in a manner that provides equal representation to all levels of each design parameter using less design points.

The $Star$ design is one of two kinds of fractional factorial design approaches. It requires the minimum number of simulations to build the regression model—$k + 1$ simulations to be done with $k$ parameters, but there are no degrees of freedom to estimate how well the full model is obeyed. The full model means that each parameter contributes significantly to the model, and there are $k$ parameters in the final regression model. To overcome this problem, another design approach is necessary. Table 5.5 shows an orthogonal array $L_8$ with four columns and two factor levels. We use this design method in our research.

**Table 5.5** $L_8$ Orthogonal Array

| No. | $x_1$ | $x_2$ | $x_3$ | $x_4$ |
|---|---|---|---|---|
| 1 | -1 | -1 | -1 | -1 |
| 2 | -1 | -1 | 1 | 1 |
| 3 | -1 | 1 | -1 | 1 |
| 4 | -1 | 1 | 1 | -1 |
| 5 | 1 | -1 | -1 | 1 |
| 6 | 1 | -1 | 1 | -1 |
| 7 | 1 | 1 | -1 | -1 |
| 8 | 1 | 1 | 1 | 1 |

Using two-level orthogonal arrays is based on the assumption that the multiple linear regression model without interaction is adequate to express the system behavior within the variation of each design parameter. While the system response is sometimes a curved surface, three-level orthogonal arrays are needed. The detailed applications of fractional factorial design method with two-level and three-level orthogonal arrays are presented in the following sections.

## 5.2 Optimization Verification

The optimization verification is to verify the correctness of the optimal design result. Based on the optimization procedure shown in Figure 5.12, there are two approaches for the optimization verification.

- Subjective Verification
- Objective Verification

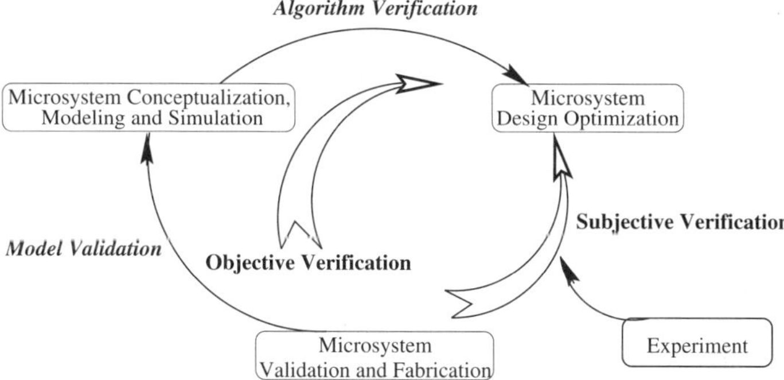


**FIGURE 5.12**
**Optimization procedure. Verification can be done either by experiments (subjective verification), or by model validation and algorithm verification (objective verification)**

### 5.2.1 Subjective Verification

The subjective verification approach verifies the optimal results directly based on existing product samples. This approach is independent of the system model and the optimization algorithm, therefore, it is also called "independent verification and validation (IV&V)" [93]. Rapid prototyping is one of the most potential subjective verification approaches [94]. It can rapidly provide a sample system based on the optimal design solution, then verify the optimal design result. This verification method can give accurate verification results; however, it is time-consuming, expensive, and sometimes the experimental data is hard to obtain. In addition, it does not provide any information for the improvement of the system model and the optimization algorithm.

### 5.2.2 Objective Verification

In objective verification, optimization results are verified based on the original design optimization procedure. Since the optimization results depend on the accuracy of the system model and the optimization algorithm, this verification approach includes two steps (Figure 5.12):

- Model Validation
  This is defined as "ensuring that a computerized model within its domain of applicability possesses a satisfactory range of accuracy consistent with the intended application of the model" [95].

- Algorithm Verification
  This is defined as "ensuring that the algorithm and its implementation are correct" [95].

#### 5.2.2.1 MEFS Model Validation

A commonly used model validation technique is the comparison validation method. Three-dimensional numerical models developed using finite element method (FEM) are typically used to verify MEFS behavioral models with analytical expressions including differential and algebraic equations (DAEs). For instance, Figure 5.13 shows a microvalve analytical behavioral model. A more detailed microvalve numerical model based on FEM is shown in Figure 5.14. The validity of the microvalve behavior model is verified by comparing the simulation results between these two models.

MEFS multi-parameter design requires model validation with respect to each design parameter. This means that the model is considered valid only if it is accurate within the scope of the variation of each design parameter, because a model may be valid for one set of experimental conditions but invalid for another. However, complete verification is costly and time-consuming. Figure 5.15 [93] shows the relationship between cost of model validation and the value of the model to the user as a function of model confidence. The cost of model validation is usually quite significant, particularly when high model confidence is required.

Several model verification techniques have been progressed in the literature [96, 31]; unfortunately, there is no set of specific tests that can easily be applied to determine the "correctness" of the model. MEFS models we used in our thesis, microvalve, micropump, and micromechanical resonator, have been verified by several research effects [16, 4, 83] based on a single design parameter. We assume that the accuracy of these models is within the scope for each design parameter, from the nominal design point to the optimal design point.

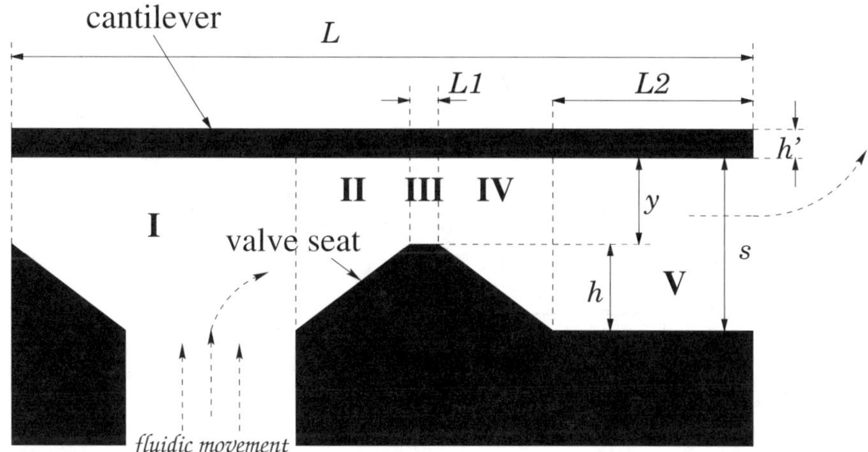


**FIGURE 5.13**
**An analytical behavioral model of an open valve of a micropump.**

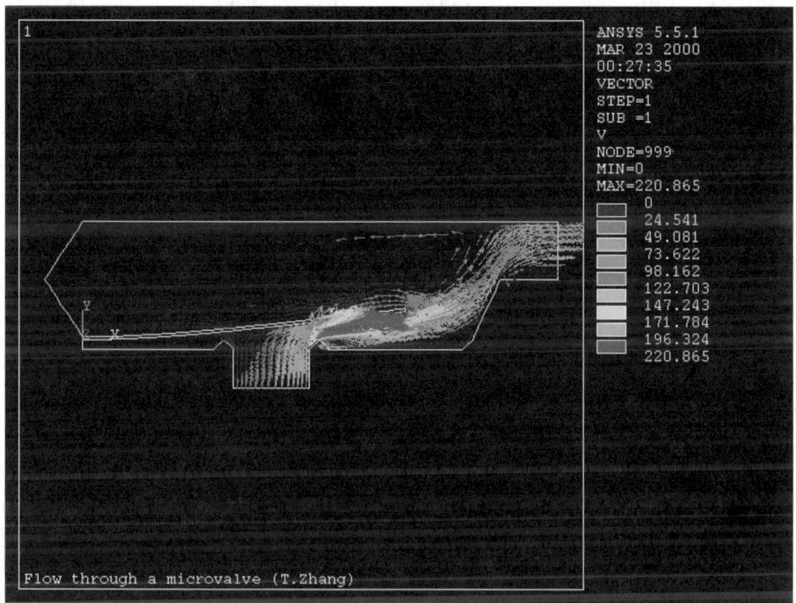


**FIGURE 5.14**
**A numerical model and simulation results of a microvalve. The model is built based on FEM using ANSYS.**

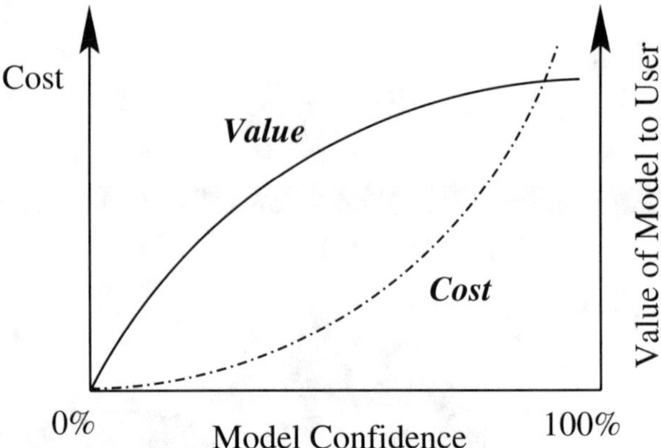


**FIGURE 5.15**
**Model confidence versus cost**

#### 5.2.2.2 Verification of Optimization Algorithms

Validation of MEFS optimization algorithms is critical to ensure that the optimization is useful. This validation can always be done by comparing the system performance between the nominal design solution and the optimal design solution. For example, in order to check the correction of the robust design of the system, the system robustness of the optimal design point can be compared with that of the nominal design point [97]. If the comparison shows the better robustness on the optimal design point, that verifies the correctness of the robust design algorithm. The comparison can be done based on the statistical analysis methodology. The detailed verification approach is presented in the following sections.

## 5.3 On-target Design Optimization

The initial design usually does not match the system design performance requirement. A design optimization process is necessary to search an on-target design solution that makes the system satisfy the design requirement. This searching process may or may not be based on the existing design solution. With increasing design complexity of composite microsystems, the search for a on-target design point in the large solution space is becoming very difficult. Therefore, it is necessary to find the most efficient direction of search (gradient) from the initial design point to the optimal design point. This objective requires a study of statistical response analyses that seek to characterize the relationships between the basic electrical/mechanical parameters and system performance. These relationships not only benefit the on-target search process [31], but they also help to improve manufacturing yield and product robustness [98]. Based on statistical response analyses, system performance variability is studied within a region around a selected design point. This allows a designer to understand how fluctuations in design parameters shift the design point and the associated system behavior. This analysis indicates the direction of search for the on-target performance design point and determines the factors that need to be more carefully analyzed during design and more carefully controlled during manufacturing. Conversely, secondary factors can be eliminated, simplifying component models, and manufacturing process control.

Performance data are a primary requirement for statistical analysis. High-level behavioral models with differential algebraic equations (DAEs) are used to generate the statistical data. These models must be validated with FEM simulations. Some of these DAEs are inherently non-linear and coupled, and they can only be solved numerically. Therefore it is not feasible to compute performance metrics and sensitivities directly from these equations. Statistical methods are thus useful in these cases. We build and simulate these behavior models based on a MEFS hierarchical modeling and simulation environment with SystemC. The steps of an algorithm for on-target design with statistical response analyses are given below.

1. Create behavioral coupled-energy models with SystemC.
2. Define the initial design point.
3. Define simulation parameters.
   (a) Identify factors and factor levels to be used in simulation for evaluating performance variance.
   (b) Construct orthogonal array of factors and levels to identify fractional factorial factor level combinations defining the required simulations.
4. Perform simulations with SystemC models (obtain simulation results).
5. Build a linear regression model.
   (a) Verify the model until the final regression model correctly presents the relationships between the system performance and the basic design parameters.
6. Obtain gradient vector for searching an on-target design point.
7. Repeat steps 3 to 6 to reach the final target point.
8. Build linear regression model at the target point.
   (a) Repeat step 3 to 5 to obtain the regression model.

9. Data Analysis

   (a) Sensitivity analysis for each design parameters.

In this section, multivariate linear regression modeling for design and process optimization is illustrated for a microvalve, which serves as a sample MEFS device. In Section 5.3.1, the fundamentals of multivariate linear regression modeling are introduced. Based on the verified circuit-level behavior model with SystemC discussed in Section 4.4, the sample performance data have been generated by using factorial design methods. Section 5.3.2 presents the statistical modeling of the microvalve. The path searching algorithm for on-target design is presented in Section 5.3.3, and the sensitivity of each design parameter is analyzed in Section 5.3.4.

## 5.3.1 Statistical Modeling and Response Analyses

In general, response performance can be related to several design parameters, called predictor variables or regressors. The number of regressors is presented by $k$. This can be done using multivariate regression analysis. Multivariate regression analysis characterizes the relationships between independent and dependent variables.

### 5.3.1.1 Multiple Linear Regression Model

Linear multivariate regression analysis assumes a regression equation of the form given in (5.18) [89].

$$y = \beta_0 + \beta_1 x_1 + \beta_2 x_2 + \cdots + \beta_k x_k + \epsilon \tag{5.18}$$

The constants $\beta_0, \beta_1, \cdots, \beta_k$, called regression coefficients, measure the expected change in the dependent variable $y$ per unit change in the associated independent variables $x_0, x_1, \cdots, x_k$. The term $\epsilon$, called the residual term or error term, accounts for variability in $y$ that cannot be accounted for by a strictly linear approximation.

Given $n$ $(n > k)$ statistical samples (simulation/experiments), the data set shown in Table 5.6 is obtained. Regressor variable values are denoted by $x_{ij}$, with $j$ identifying the input parameter and $i$ identifying the experiment. These regressor variables, $x_{ij}$, cover an area that is defined as the experiment area. The response performance variables are denoted by $y_1, y_2, \cdots, y_n$.

Using the data in Table 5.6, the multivariate linear regression model can be rewritten in the form

$$\boldsymbol{y} = \boldsymbol{X\beta} + \boldsymbol{\epsilon} \tag{5.19}$$

Where $\boldsymbol{y} = (y_1, y_2, \cdots, y_n)^T$ is an $(n \times 1)$ vector of the system performance observations, $\boldsymbol{\beta} = (\beta_0, \beta_1, \cdots, \beta_k)^T$ is a $((k+1) \times 1)$ vector of the partial regression

**Table 5.6** Simulation Data for Multiple Regression Analysis

| Performance | Design Parameters | | | | |
|---|---|---|---|---|---|
| $y$ | $x_1$ | $x_2$ | $x_3$ | $\cdots$ | $x_k$ |
| $y_1$ | $x_{11}$ | $x_{12}$ | $x_{13}$ | $\cdots$ | $x_{1k}$ |
| $y_2$ | $x_{21}$ | $x_{22}$ | $x_{23}$ | $\cdots$ | $x_{2k}$ |
| . | . | . | . | . | . |
| . | . | . | . | . | . |
| . | . | . | . | . | . |
| $y_n$ | $x_{n1}$ | $x_{n2}$ | $x_{n3}$ | $\cdots$ | $x_{nk}$ |

coefficients, and $\epsilon = (\epsilon_1, \epsilon_2, \cdots, \epsilon_n)^T$ is an $(n \times 1)$ vector of the residuals. $\boldsymbol{X}$ is an $n \times (k+1)$ matrix of the values of the independent design parameters.

$$\boldsymbol{X} = \begin{bmatrix} 1 & x_{11} & x_{12} & \cdots & x_{1k} \\ 1 & x_{21} & x_{22} & \cdots & x_{2k} \\ \cdot & \cdot & \cdot & & \cdot \\ \cdot & \cdot & \cdot & & \cdot \\ \cdot & \cdot & \cdot & & \cdot \\ 1 & x_{n1} & x_{n2} & \cdots & x_{nk} \end{bmatrix} \tag{5.20}$$

With the least square estimation, the matrix form of the least square estimators of partial regression coefficients $\boldsymbol{\beta}$ is given by [89]

$$\boldsymbol{b} = (\boldsymbol{X}^T\boldsymbol{X})^{-1}\boldsymbol{X}^T\boldsymbol{y} \tag{5.21}$$

and the multivariate linear regression equation for the fitted value $\hat{\boldsymbol{y}}$ is given by

$$\hat{\boldsymbol{y}} = \boldsymbol{X}\boldsymbol{b} \tag{5.22}$$

#### 5.3.1.2 Verification of Statistical Models

Verification of statistical models involves evaluating the quality of the model in representing the sample data. Quality refers to both regression error and its statistical significance.

- Test for Efficiency of Regression
  A measure of efficiency of a regression model is given by the coefficient of determination, $R^2$. The coefficient of determination is defined as the ratio of the regression sum of squares ($SSR$) to the sum of squares ($SYY$), and is given by [31]

$$R^2 = \frac{SSR}{SYY} \tag{5.23}$$

where, $SYY = \sum_{i=1}^{n} y_i^2 - (\sum_{i=1}^{n} y_i)^2$, and $SSR = \boldsymbol{b}'\boldsymbol{X}'\boldsymbol{y} - (\sum_{i=1}^{n} y_i)^2/n$. To correct for dependencies on the number of regressor variables that may statistically have no effect on regression efficiency, an adjusted coefficient of determination is often used

$$R_{adj}^2 = 1 - (\frac{n-1}{n-k-1})(1-R^2) \tag{5.24}$$

where, $n$ is the sample size and $k$ is the number of regressor variables. The coefficient of determination has the range $0 \leq R^2 \leq 1$, with 0 denoting no regression efficiency over using the regression model and 1 denoting a perfect linear fit. The square root of the coefficient of determination is the coefficient of correlation.

- Test for Significance of Individual Regression Coefficients

  Since (5.21) is an estimate of the regression coefficients, $\boldsymbol{\beta}$, there is a need to test the statistical significance of each regression coefficient. This test is useful in determining which regressor variable must be included/excluded in the multivariate linear regression.

  The ratio of the estimated regression coefficient to an estimate of its standard error given by (5.25) follows a $t$ distribution [31], i.e.

$$t_{H_0} = \frac{b_j}{s_{b_j}} = \frac{b_j}{\sqrt{\hat{\sigma_e}^2 c_{jj}}} \tag{5.25}$$

  where, $\sqrt{\hat{\sigma_e}^2}$ is the estimate of the standard deviation of residuals, also called the standard error of estimate [31].

$$\sqrt{\hat{\sigma}^2} = s_e = \sqrt{MS_E} = \sqrt{(\boldsymbol{y}^T\boldsymbol{y} - \boldsymbol{b}^T\boldsymbol{X}^T\boldsymbol{y})/(n-k-1)}$$

  The term $c_{jj}$ is the diagonal element of $(\boldsymbol{X}^T\boldsymbol{X})^{-1}$ corresponding to $b_j$. If $|t_{H_0}|$ exceeds $t_{\alpha/2,n-k-1}$, the regression coefficient is statistically greater than zero, and the related regressor variable should be included in the multivariate linear regression model. The $100(1-\alpha)\%$ confidence interval for the regression coefficients, $\beta_j, j = 0, 1, 2, \cdots, k$, follows directly from (5.25) and is given by [31]

$$b_j - (t_{\alpha/2,n-k-1})(s_{b_j}) \leq \beta_j \leq b_j + (t_{\alpha/2,n-k-1})(s_{b_j}) \tag{5.26}$$

  In repeated statistical response analyses using data samples of size $n$, the confidence interval will contain the true (population) regression coefficient $100(1-\alpha)\%$ of the time. The $100(1-\alpha)\%$ confidence interval can also be computed for the estimate $\hat{y}$ of the mean of $y$ for a given set of regressor variable values, $\boldsymbol{X}_\gamma = [1, x_{\gamma 1}, x_{\gamma 2}, x_{\gamma 3}, \cdots, x_{\gamma k}]$.

$$\hat{y}(X_\gamma) - (t_{\alpha/2,n-k-1})(s_{\hat{y}}) \leq y_\gamma \leq \hat{y}(X_\gamma) + (t_{\alpha/2,n-k-1})(s_{\hat{y}}) \tag{5.27}$$

  where $s_{\hat{y}} = \sqrt{\hat{\sigma_e}^2 \boldsymbol{X}_\gamma (\boldsymbol{X}^T\boldsymbol{X})^{-1}\boldsymbol{X}_\gamma^T}$.

- Outliers Analysis
  Scaling algorithm generating studentized residuals is given by (5.28)

$$r_i = \frac{e_i}{\sqrt{\hat{\sigma_e}^2(1 - h_{ii})}}, \qquad i = 1, 2, \cdots, n \tag{5.28}$$

Again, $\sigma_e^2$ is the estimate of residual variance. The $e_i$ is the $i^{\text{th}}$ element in $\boldsymbol{e}$, i.e. $\boldsymbol{e} = \boldsymbol{y} - \hat{\boldsymbol{y}}$. The term $h_{ii}$ is the $i^{\text{th}}$ diagonal element of the $\boldsymbol{H}$ matrix, which is the projection matrix of $y$, with

$$\boldsymbol{H} = \boldsymbol{X}(\boldsymbol{X}^T\boldsymbol{X})^{-1}\boldsymbol{X}^T$$

The studentized residuals have zero mean and approximately unity variance. It is useful when looking for **outliers**, whose studentized residual value does not lie in the interval $-3 \leq r_i \leq 3$. These outliers must be carefully examined. They may denote the fact that there is unusual change in the response surface, and the fitted model is a poor approximation of the true response surface.

### 5.3.2 Statistical Modeling of a Microvalve

The verified circuit-level behavior model of a microvalve as discussed in Section 4.4 can be used for statistical response analyses. The design parameters affecting microvalve performance and their nominal design values are shown in Table 5.7. These basic parameters are subject to disturbances due to the fabrication processes. Their tolerances are assumed in Table 5.8. Again, tolerances are assumed to be constant within the design envelope.

**Table 5.7** Elemental Parameters and Initial Nominal Design Values

| Parameters | Values | Units |
|---|---|---|
| Length of the cantilever ($L$) | 1600 | $\mu m$ |
| Width of the cantilever($b'$) | 1000 | $\mu m$ |
| Thickness of the cantilever ($h'$) | 15 | $\mu m$ |
| Height of the valve seat ($h$) | 50 | $\mu m$ |
| Length of the valve seat ($l_1$) | 5 | $\mu m$ |
| Width of the valve seat ($b$) | 400 | $\mu m$ |
| Length of the cantilever over valve seat($l_2$) | 100 | $\mu m$ |
| Young's Modulus ($E$) | 146.9 | GPa |
| Air Pressure($p$) | 100000 | $Pa$ |

**Table 5.8** Noise Factors for Microvalve

| Noise Factors | Tolerance Level (-1) | (0) | (+1) | Unit |
|---|---|---|---|---|
| $L$ | -0.2 | 0 | 0.2 | $\mu m$ |
| $b'$ | -0.2 | 0 | 0.2 | $\mu m$ |
| $h'$ | -0.2 | 0 | 0.2 | $\mu m$ |
| $l_1$ | -0.2 | 0 | 0.2 | $\mu m$ |
| $b$ | -0.2 | 0 | 0.2 | $\mu m$ |
| $h$ | -0.2 | 0 | 0.2 | $\mu m$ |
| $l_2$ | -0.2 | 0 | 0.2 | $\mu m$ |
| $p$ | -1 | 0 | 1 | Pa |

The Young modulus $E$ is related to the design material, and is assumed to maintain constant. Instead of a complete factorial design, a standard orthogonal array $L_{16}$ with 8 columns and two factor levels is used to define sampled circuit simulations. The $L_{16}$ orthogonal array and simulation results for the microvalve performance metrics of flow rate $\Phi$ are given in Table 5.9.

**Table 5.9** Noise Factor Simulation Design and System Response

| No. | $L$ | $b'$ | $h'$ | $h$ | $l_1$ | $b$ | $l_2$ | $p$ | Flowrate $(ml/min)$ |
|---|---|---|---|---|---|---|---|---|---|
| 1 | -1 | -1 | -1 | -1 | -1 | -1 | -1 | -1 | 5.8570 |
| 2 | -1 | -1 | 1 | -1 | 1 | 1 | 1 | -1 | 5.8619 |
| 3 | -1 | 1 | -1 | 1 | -1 | 1 | 1 | -1 | 5.8652 |
| 4 | -1 | 1 | 1 | 1 | 1 | -1 | -1 | -1 | 5.8584 |
| 5 | 1 | -1 | -1 | 1 | 1 | -1 | 1 | -1 | 5.8585 |
| 6 | 1 | -1 | 1 | 1 | -1 | 1 | -1 | -1 | 5.8634 |
| 7 | 1 | 1 | -1 | -1 | 1 | 1 | -1 | -1 | 5.8667 |
| 8 | 1 | 1 | 1 | -1 | -1 | -1 | 1 | -1 | 5.8599 |
| 9 | 1 | 1 | 1 | 1 | 1 | 1 | 1 | 1 | 5.8660 |
| 10 | 1 | 1 | -1 | 1 | -1 | -1 | -1 | 1 | 5.8610 |
| 11 | 1 | -1 | 1 | -1 | 1 | -1 | -1 | 1 | 5.8577 |
| 12 | 1 | -1 | -1 | -1 | -1 | 1 | 1 | 1 | 5.8645 |
| 13 | -1 | 1 | 1 | -1 | -1 | 1 | -1 | 1 | 5.8645 |
| 14 | -1 | 1 | -1 | -1 | 1 | -1 | 1 | 1 | 5.8595 |
| 15 | -1 | -1 | 1 | 1 | -1 | -1 | 1 | 1 | 5.8563 |
| 16 | -1 | -1 | -1 | 1 | 1 | 1 | -1 | 1 | 5.8631 |

The regression model of microvalve flow rate $\Phi$ is shown in (5.29). It is obtained using the multivariate linear regression techniques discussed in previous sections.

$$\hat{\Phi} = -20.4734 + 0.0037\,L + 0.0058\,b' - 0.0023\,h' + 0.0001\,h + 0.0147\,b + 0.0004\,p \quad (5.29)$$

The coefficient of determination $R^2$ and the adjusted coefficient of determination $R^2_{adj}$ statistics show a very good fit, since

$$R^2 = 1.000, \quad R^2_{adj} = 1.000$$

Table 5.10 shows significance testing for individual regression coefficients. Assuming $\alpha = 0.05$, $t$-statistics of these parameters are larger than the $t$-criterion $t_{0.025,9} = 2.262$, verifying that these design parameters contribute significantly to the regression model.

**Table 5.10** Test for Individual Regression Coefficients for Sensitivity Model

| $t$ Distribution | Regression Coefficients | | | | | |
|---|---|---|---|---|---|---|
| | $\beta_1\ (L)$ | $\beta_2\ (b')$ | $\beta_3\ (h')$ | $\beta_4\ (h)$ | $\beta_5\ (b)$ | $\beta_6\ (p)$ |
| $t_{H_0}$ | 2563 | 4077 | 1602 | 46 | 10228 | 307 |

The confidence intervals for the individual regression coefficients with $\alpha = 0.05$ are shown in Table 5.11.

**Table 5.11** The 95% Confidence Interval for Individual Regression Coefficients

| Design Parameters | Range $\pm(t_{\alpha/2,n-k-1})(s_{b_j})$ | Confidence Intervals |
|---|---|---|
| $L$ | $\pm 0.6482e-6$ | $0.3699e-2 \leq \beta_1 \leq 0.3706e-2$ |
| $b'$ | $\pm 0.6482e-6$ | $0.5794e-2 \leq \beta_2 \leq 0.5806e-2$ |
| $h'$ | $\pm 0.6482e-6$ | $-0.2306e-2 \leq \beta_3 \leq -0.2294e-2$ |
| $h$ | $\pm 0.6482e-6$ | $0.0994e-2 \leq \beta_4 \leq 0.1006e-2$ |
| $b$ | $\pm 0.6482e-6$ | $0.1464e-2 \leq \beta_5 \leq 0.1471e-2$ |
| $p$ | $\pm 0.6482e-6$ | $0.0994e-2 \leq \beta_6 \leq 0.1006e-2$ |

Figure 5.16 compares the microvalve flow rate obtained from the simulation, represented by (+), and that from regression model, represented by (o). The standardized residual $r$ is within the acceptable range $[-1.4541, 1.4541]$. It shows a good match between simulation data and the regression model.

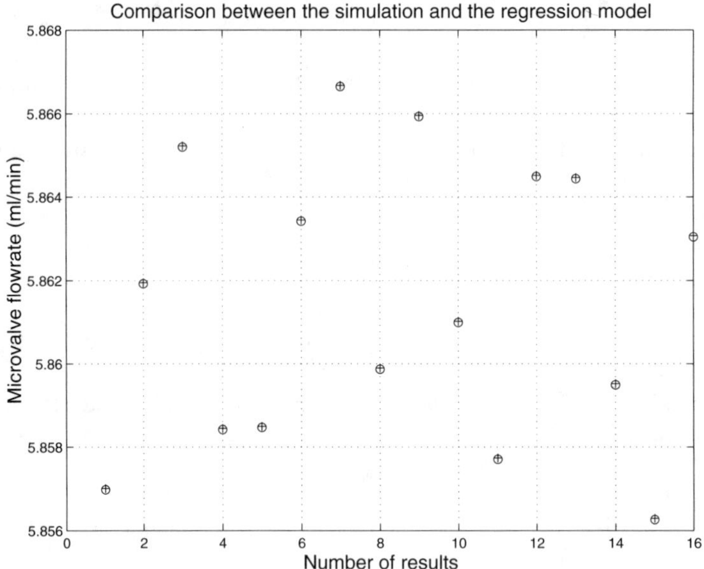


**FIGURE 5.16**
**Comparison between results (+) by simulation data and results (o) by multiple regression model shows a perfect match.**

### 5.3.3 Search for On-target Design Point

The performance on the initial design point may not always match the design requirements. In such cases, the search for a new design point that matches the performance requirements of the system is necessary. The method of steepest ascent/descent [31], which maximizes the increase or decrease in system performance along a searching path, can be used to find an optimal response point.

Using the regression model assumed in (5.18), the variation of the design variable $x_i$ along the path of steepest ascent is proportional to the magnitude of the regression coefficient, $b_i$, with the direction taken being the sign of the coefficient. A steepest descent search requires the direction to be opposite to the sign of the coefficient.

Table 5.12 lists the steepest ascent/descent increment $\triangle$ for the main design parameters. These numbers are representative of the gradient vector from the initial design point to the performance on-target solution. The searching direction may be changed depending on the regression model built along the on-target searching path. In addition, the searching path is also characterized by a confidence region, which determines how accurately a path is being estimated. Table 5.11 shows the confidence intervals for the individual regression coefficients, which is a measure of the accuracy of the coefficients. In order to obtain accurate gradient vectors for optimal on-target system search, the *Range* indicated in Table 5.11 has to be as small as possible for a good confidence interval.

**Table 5.12** Steepest Ascent/descent Increment

| Design Parameters | $L$ | $b'$ | $h'$ | $h$ | $b$ | $p$ |
|---|---|---|---|---|---|---|
| Increment $\triangle$ | 1.600 | 2.545 | -1 | 0.029 | 6.385 | 0.038 |

Usually, the increment along the path is based on a movement of one particular variable. In order to avoid infeasible increments of the design parameters, a variable from the regression model is selected as the basis for the increment, whose design value is minimal. For example, $h'$ is selected at the nominal design point. The steps of the searching algorithm with the adjustable increment are defined as the following:

1. Compute a path of steepest ascent if the target performance is larger than the performance on the initial design point. If the target performance is less than the performance on the initial design point, compute the path of steepest descent.

2. Conduct simulations along the path. That is, do single or replicated runs and observe the response value. The results should normally show the improvement of performance.

3. At some region along the path, the improvement will decline or reverse. Then, the new design before reversal should replace the initial design point. Another first-order model fitting is carried out, and the new searching direction is created. If the new design point possesses the maximum value of performance (ascent design), or the minimum value of performance (descent design), it means that there is no design solution in this local design space. The searching algorithm stops, and another new initial design point is necessary.

4. At some region along the path, the improvement will pass the performance target, such as the run 7 in Table 5.13. The new design point will replace the original nominal design point. In addition, the direction of the increment is reversed, and the increment is reduced by half. Go back to step 2.

5. After reaching the optimal design point, another first-order model fitting is carried out. A test of lack of fit is made, and the sensitivity analysis for each parameter is conducted.

At the nominal design point (Table 5.7), the microvalve flow rate is $5.86ml/min$. Table 5.13 shows the appropriates coordinate along the searching path in the design parameters. The target flow rate is $6.50ml/min$. After 10 searching steps, the system performance matches the design target. The increment has been changed 4 times. Each block in Table 5.13 shows these changes.

**Table 5.13** Coordinates on the Searching Path of Steepest Ascent in Design Parameters

| Run | $L$ | $b'$ | $h'$ | $h$ | $l_1$ | $b$ | $l_2$ | $p$ | $\Phi$ |
|---|---|---|---|---|---|---|---|---|---|
| 1) Base | 1600 | 1000 | 15 | 50 | 5 | 400 | 100 | 100000 | 5.86 |
| $\hat{\Phi} = -20.4734 + 0.0037L + 0.0058b' - 0.0023h' + 0.0001h + 0.0147b + 0.0004p$ | | | | | | | | | |
| $\triangle$ | 1.60 | 2.55 | -1 | 0.03 | 0 | 6.39 | 0 | 0.04 | |
| 2) Base + $\triangle$ | 1601.6 | 1002.55 | 14 | 50.03 | 5 | 406.39 | 100 | 100000 | 5.98 |
| 3) Base + 2$\triangle$ | 1603.2 | 1005.1 | 13 | 50.06 | 5 | 412.78 | 100 | 100000 | 6.09 |
| 4) Base + 3$\triangle$ | 1604.8 | 1007.6 | 12 | 50.09 | 5 | 419.15 | 100 | 100000 | 6.21 |
| 5) Base + 4$\triangle$ | 1606.4 | 1010.2 | 11 | 50.11 | 5 | 425.54 | 100 | 100000 | 6.32 |
| 6) Base + 5$\triangle$ | 1608 | 1012.7 | 10 | 50.14 | 5 | 431.92 | 100 | 100000 | 6.42 |
| 7) Base + 6$\triangle$ | 1609.6 | 1015.3 | 9 | 50.17 | 5 | 438.31 | 100 | 100000 | 6.51 |
| Base | 1609.6 | 1015.3 | 9 | 50.17 | 5 | 438.31 | 100 | 100000 | 6.51 |
| $\triangle' = \triangle/2$ | 0.8 | 1.27 | -0.5 | 0.01 | 0 | 3.19 | 0 | 0.02 | |
| 8) Base - $\triangle'$ | 1608.8 | 1014 | 9.5 | 50.16 | 5 | 435.12 | 100 | 100000 | 6.47 |
| Base | 1608.8 | 1014 | 9.5 | 50.16 | 5 | 435.12 | 100 | 100000 | 6.47 |
| $\triangle'' = \triangle'/2$ | 0.4 | 0.64 | -0.25 | 0.01 | 0 | 1.6 | 0 | 0.01 | |
| 9) Base + $\triangle''$ | 1609.2 | 1014.6 | 9.25 | 50.16 | 5 | 436.71 | 100 | 100000 | 6.49 |
| Base | 1609.2 | 1014.6 | 9.25 | 50.16 | 5 | 436.71 | 100 | 100000 | 6.49 |
| $\triangle''' = \triangle''/2$ | 0.2 | 0.32 | -0.13 | 0.01 | 0 | 0.8 | 0 | 0.01 | |
| 10) Base + $\triangle'''$ | 1609.4 | 1015 | 9.12 | 50.17 | 5 | 437.51 | 100 | 100000 | 6.50 |
| $\hat{\Phi} = -14.3406 + 0.0046\,L + 0.0065\,b' + 0.034\,h' + 0.0149\,b$ | | | | | | | | | |

### 5.3.4 Sensitivity Analysis

A new regression model of microvalve flow rate at the optimal design point is shown in (5.30).

$$\hat{\Phi} = -14.3406 + 0.0046\,L + 0.0065\,b' + 0.034\,h' + 0.0149\,b \tag{5.30}$$

The coefficient of determination $R^2$ and the adjusted coefficient of determination $R^2_{adj}$ statistics show a very good fit.

$$R^2 = 0.9994, \quad R^2_{adj} = 0.9991$$

As shown in (5.31), the main effect $\rho_i$ of a regression variable $x_i$ is the average change of system response when the variable $x_i$ fluctuates from its low-level value to its high-level value, and all other variables hold constant. Thus, main effect denotes the importance of a design parameter to system performance.

$$\rho_i = \frac{\partial \Phi}{\partial x_1} \Delta x_i \tag{5.31}$$

Having verified the multivariate linear regression model given in (5.30), the related main effect analysis and resulting overall sensitivity ranking for the regression coefficients are shown in Table 5.14, which indicates that the thickness of the cantilever beam $h'$ should be carefully controlled to improve fabrication yield and system robustness.

**Table 5.14** Main Effect Analysis for Sensitivity Model

| Design Parameters | $L$ | $b'$ | $h'$ | $b$ |
|---|---|---|---|---|
| Main Effect | 9.2e-4 | 13.1e-4 | 68e-4 | 29.7e-4 |
| Sensitivity Ranking | 4 | 3 | 1 | 2 |

Statistical modeling and response analyses are useful for developing macromodels for composite microsystems. (5.29) and (5.30) represent complex physical models with multivariate linear regression models, which can be used for efficient design space exploration. By using the statistical response analysis, issues of designing for on-target response, manufacturing yield and operational robustness can be addressed.

## 5.4 Robust Design Optimization[1]

With the number of pilot applications of integrated composite microsystem growing, there is a need for robust design optimizations to support all aspects of product development, including design, manufacturing, and operational use. New designs that seek enhanced performance and function require an understanding of the performance space and design issues. Studying statistical parametric performance relations can aid in understanding the limits of present transduction mechanisms and explore new transduction mechanisms. Due to the high levels of environmental sensitivity and small dimensions associated with integrated circuit fabrication, manufacturing of composite microsystems is increasingly important in determining commercial viability and affordability. Manufacturing yields are presently significantly lower for MEMS and MEFS than comparable microelectronics [99]. Finally, operational robustness requires understanding the environmental limits of transduction and the effect on microsystem performance. The limitation is associated with component aging and degradation that may cause unwanted drifts in nominal design settings. Thus, there is a need for detailed and systematic analyses of the relationships between basic electrical/mechanical design parameters and overall design function and performance.

Taguchi experimental design and statistical process control methods [30] provide efficient means for conducting performance variability reduction and parametric sensitivity analyses. They are used for off-line parametric optimization control and high performance design. The objective here is to identify the parameters or factors most influential in determining a performance metric, and to compute the settings of the parameters that yield both an acceptable performance metric and minimize the influence of parametric variations. In other words, by choosing levels of design parameters, sensitivity to product design parameter variation due to the variance of a set of noise factors is reduced. Experimental data is generated using the fractional factorial design methodology.

Taguchi performance variability reduction and parametric sensitivity analyses introduce design-for-manufacturing into behavioral modeling. Conducting off-line control modeling and simulation before actual fabrication is becoming an important aspect for composite microsystem design optimization, because of the high costs of fabrication and the difficulties of controlling the inherent stochastic nature of fabrication processes. The emphasis is on maximizing the insensitivity of the design

[1]This section is based in part on "A. Dewey, H. Ren, and T. Zhang, Behavior Modeling of Microelectromechanical Systems (MEMS) with Statistical Performance Variability Reduction and Sensitivity Analysis. IEEE Transaction on Circuit and Systems II : Analog and Digital Signal Processing, vol. 47, no. 2, pp. 105-113, Feb. 2000." © 2000 IEEE. Reprinted by permission.

to a known set of parametric variations, rather than on minimizing the parametric variations themselves.

The organization of this section is as follows. Section 5.4.1 presents the formulation of statistical noise sensitivity reduction for optimizing design performance stability, followed by sensitivity or variance analysis to determine and rank-order the contributions of the individual parameters to the optimized design performance stability. Based on the behavioral models of a electrostatic-comb microresonator presented in Section 5.1, Section 5.4.2 presents the results for Taguchi analyses.

## 5.4.1 Statistical Response Analysis

Statistical modeling and analyses, which can be profitably used to understand the limits of present transduction mechanisms and explore new transduction, is crucial to continued advancements in design, producting yield, scalability limits, and robustness.

Due to the variance of elemental parameters, overall product performance fluctuates around a nominal design point. Excessive fluctuation moves the nominal design point outside of acceptable tolerance limits and results in unacceptable performance. Hence, it is desirable to be able to select optimum settings or values for design parameters such that the product is functional, exhibits a high level of performance under a wide range of conditions, and is robust against factors causing variability [87]. This objective can be achieved by conducting statistical response analysis involving two aspects: design optimization with performance variability reduction, and manufacturing/operating environment optimization control with sensitivity/variance analysis.

### 5.4.1.1 Performance Variability Reduction

Figure 5.17 illustrates the concept of performance variability or noise sensitivity reduction.

Design parameters are denoted by $x_1$ and $x_2$. Relationships between variability in design parameters $x_1$ and $x_2$ and corresponding variability in system performance are shown by the lefthand and righthand graphs, respectively. Due to the nonlinear relationship between the design parameter $x_1$ and system performance response $f$, the change of design point from $x_{11}$ to $x_{12}$ results in a performance variability reduction. A similar change in design point from $x_{21}$ to $x_{22}$ yields no performance variability reduction due to the linear relationship between design parameter $x_2$ and system performance response $f$.

Performance variability reduction concerns identifying design parameters that have the most influence on performance variability. It also concerns setting the values of

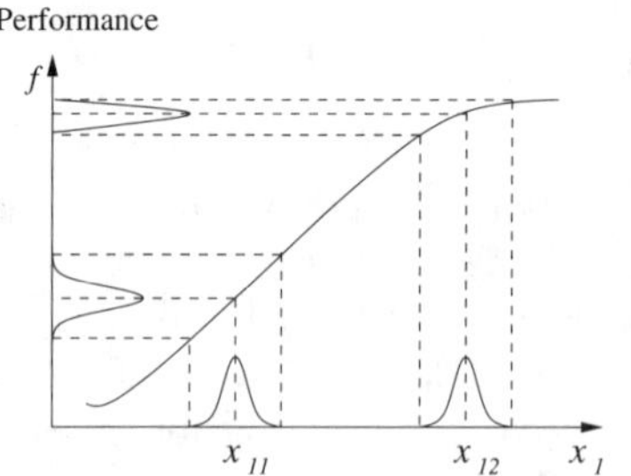


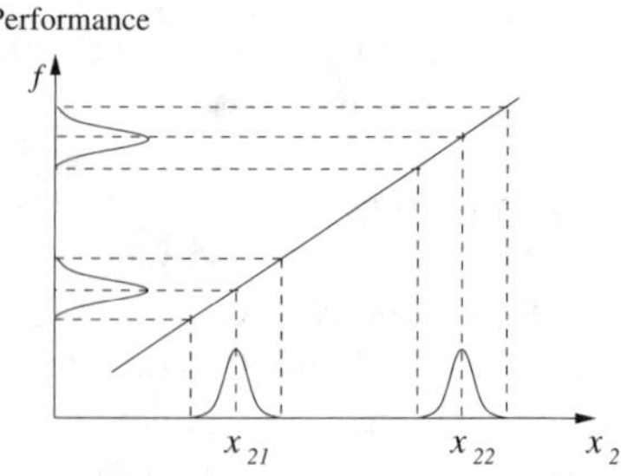


**FIGURE 5.17**
**The way to performance variability reduction. The figure on the left shows the non-linear relationship between a design parameter and the system performance. The figure on the right shows the linear relationship between the system performance and a design parameter**

the design parameters to move the design point into the region where performance sensitivity to parametric variations is minimized. In addition, design parameters having the least influence on performance variability are employed to perform functional tuning to ensure overall system performance meets target specifications. For example, assume in Figure 5.17 an initial design point of $(x_{11}, x_{22})$. After moving the design parameter $x_1$ from $x_{11}$ to $x_{12}$ to reduce performance variability, design parameter $x_2$ could be adjusted from value $x_{22}$ to value $x_{21}$ as a tuning factor to maintain the performance target. In this scenario, design parameter $x_1$ is used to improve variability reduction at the expense of nominal system performance. Undesirable shifts in nominal system performance are, in turn, compensated for via design parameter $x_2$.

Performance variability is computed based on a statistical metric suggested by Taguchi, signal-to-noise ratio (SNR) [30]. System performance is considered a "signal" and parametric fluctuations are considered "noise". Signal-to-noise ratio isolates parametric fluctuations from the performance mean. Three common formulations of the signal-to-noise ratio (SNR) objective function are given below:

- Minimizing the performance response.

$$SNR = -10\ \log_{10}\left[\frac{1}{n}\sum_{i=1}^{n}\frac{1}{y_i^2}\right] \tag{5.32}$$

- Maximizing the performance response.

$$SNR = -10\ \log_{10}\left[\frac{1}{n}\sum_{i=1}^{n}y_i^2\right] \tag{5.33}$$

- Target a special performance specification while minimizing performance variance.

$$SNR = -10 \log_{10} \left[ \frac{\sigma^2}{\mu^2} \right] \tag{5.34}$$

where

- $n$ is the number of simulation replicate corresponding to each setting of design parameter combination.
- $\mu$ is the mean of overall performance response: $\mu = \frac{1}{n} \sum_{i=1}^{n} y_i$
- $\sigma^2$ is the sample variance of performance response: $\sigma^2 = \frac{1}{n-1} \sum_{i=1}^{n} (y_i - \mu)^2$

$SNR$ transforms the performance response into the log domain and provides a standard representation of different design performance variability reduction objectives; the objective functions are generally constructed such that the larger the signal-to-noise ratio, the better the performance. For instance, when the design objective is to maximize a performance metric, the larger the performance response, the larger its associated $SNR$ (5.32). In a similar manner, when the design objective is to minimize a performance metric, the smaller the performance response, the larger its associated $SNR$ (5.33). When $SNR$ is applied to on-target design, the smaller the performance variance to the target, the larger the associated $SNR$ (5.34).

Performance variability reduction computes the effect of each design parameter at several settings or levels on $SNR$ and uses the results to determine the best combination of parameter settings for optimal performance stability. Parameters are called factors and a particular parameter value is called a level. Combinations of parameters and values (factor levels) are delineated using orthogonal arrays [92]. Parameters are listed horizontally, forming the columns, and experiments or combinations of values of the parameters are listed vertically, forming the rows. An orthogonal array possess the property that all columns are mutually orthogonal in that, for any pair of columns, all combinations of factor levels occur and they occur an equal number of times.

The effect of a design parameter at a particular setting (factor level), called the factor main effect, is defined as the deviation the factor level causes from the overall mean of the performance response and is given by

$$M_A^j = \frac{1}{n} \sum_{i=1}^{n} SNR_i \tag{5.35}$$

where,

- $M_A^j$ denotes the effect of factor A at level $j$ on $SNR$, and

- $n$ is the number of experiments (simulations) involving the factor A set to level $j$.

For instance, the factor main effect of factor $A$ at level $j = 3$ on $SNR$ is computed as the average of the $SNR$s corresponding to each performance response where factor $A$ is set to level $j = 3$.

Since the factor main effect represents how close the performance response caused by a factor level is to the design objective, parameter settings having the largest factor main effect are desirable. That is, levels that maximize the signal-to-noise ratio result in minimization of performance variability. When the target performance responses are taken into account, however, the actual combination of design parameter levels should be adjusted before the final selection of parameter settings. The adjustment is based on the effect of factors on $SNR$ and performance mean.

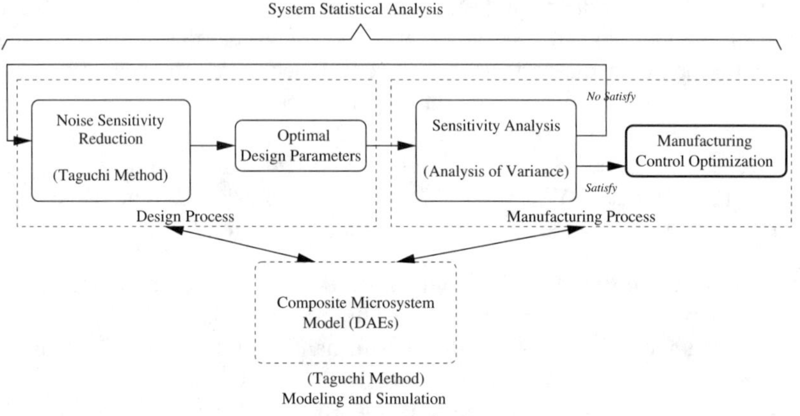


**FIGURE 5.18**
**Composite microystem statistical modeling, simulation, and analysis procedure consists of the design process and the manufacturing process.**

Figure 5.18 shows the behavioral modeling approach for robust design in more detail. The summary design and optimization process steps with an on-nominal-target design objective function are listed below:

1. Build circuit-level model
   (a) Create the verified behavioral coupled-energy circuit-level model of the system.
2. Define simulations

   (a) Identify factors and factor levels used in simulation for evaluating performance variance.

   (b) Construct orthogonal array of factors and levels to identify fractional factorial factor level combinations defining required simulations.

3. Conduct simulations

4. Optimal design analysis

   (a) Calculate the mean and variance for each factor level combination of design parameters.

   (b) Compute the influence of design parameters on $SNR$.

   (c) Compute the influence of design parameters on performance mean.

   (d) For design parameters having a significant influence on $SNR$, select levels that maximize $SNR$.

   (e) Select any design parameter(s) having negligible effect on $SNR$, but significant effect on the performance mean, as tuning factor(s) to bring the performance mean on target. The nominal value(s) of the adjustment factor(s) can be determined during design via simulation and precisely set during manufacturing via final acceptance testing/inspection.

#### 5.4.1.2 Sensitivity Analysis

After obtaining the optimum setting of design parameters to minimize overall performance variability, the second part of statistical response analysis assesses the relative contribution of each design parameter to the optimized performance variability. Relative design parameter sensitivity identifies which parameters' variations (noise factors) are limiting factors in the manufacturing process and, consequently prioritizes statistical processes monitoring and control efforts.

The sum-of-squared values [98] of performance response can be used for system sensitivity or variance analysis. Sensitivity analysis decomposes the total variability of response into separate elements representing the variance of each factor and their interaction.

Let $Y_{ij}$ denote the performance response when factor $x_1$ is set to its $i^{th}$ level and factor $x_2$ is set to its $j^{th}$ level, as shown in Table 5.15. The deviation of the performance response at a particular factor level from the overall mean performance response $(\overline{Y})$ computed with all factors set to their nominal levels is denoted by $(Y_{ij} - \overline{Y})$ and estimated by the additive effects of the following elements:

- The deviation of the mean performance response at the $i^{th}$ level of factor $x_1$ from the overall mean performance response estimated as $(\overline{Y}_{i+} - \overline{Y})$, where

**Table 5.15** Performance Response for Two Factors/Two Levels

| | | Factor $x_2$ | |
|---|---|---|---|
| | | $+1$ | $-1$ |
| Factor | $+1$ | $Y_{11}$ | $Y_{12}$ |
| $x_1$ | $-1$ | $Y_{21}$ | $Y_{22}$ |

$\overline{Y}_{i+}$ is given by

$$\overline{Y}_{i+} = \frac{1}{2}\sum_{j=1}^{2} Y_{ij}$$

- The deviation of the mean performance response at the $j^{th}$ level of factor $x_2$ from the overall mean performance response estimated as $(\overline{Y}_{+j} - \overline{Y})$, where $\overline{Y}_{+j}$ is given by

$$\overline{Y}_{+j} = \frac{1}{2}\sum_{i=1}^{2} Y_{ij}$$

- The deviation of the mean performance response at the $(i, j)^{th}$ interaction of $x_1$ and $x_2$ from the overall mean performance response estimated as $(Y_{ij} - \overline{Y}_{i+} - \overline{Y}_{+j} + \overline{Y})$.

Thus, deviation of the performance response at a particular factor level from the overall mean performance response can be expressed as

$$Y_{ij} - \overline{Y} = (\overline{Y}_{i+} - \overline{Y}) + (\overline{Y}_{+j} - \overline{Y}) + (Y_{ij} - \overline{Y}_{i+} - \overline{Y}_{+j} + \overline{Y}) \qquad (5.36)$$

Squaring both sides of (5.36) and summing over the indices $i, j$ yields the following partition of the total sum of squares $(SST)$ of the performance response for example in Table 5.15.

$$SST = SSX_1 + SSX_2 + SSX_1X_2 \qquad (5.37)$$

Where,

$$SST = \sum_{i=1}^{2}\sum_{j=1}^{2}(Y_{ij} - \overline{Y})^2 \qquad (5.38)$$

$$SSX_1 = \sum_{i=1}^{2}\sum_{j=1}^{2}(\overline{Y}_{i+} - \overline{Y})^2 = 2\sum_{i=1}^{2}(\overline{Y}_{i+} - \overline{Y})^2 \qquad (5.39)$$

$$SSX_2 = \sum_{i=1}^{2}\sum_{j=1}^{2}(\overline{Y}_{+j} - \overline{Y})^2 = 2\sum_{j=1}^{2}(\overline{Y}_{+j} - \overline{Y})^2 \qquad (5.40)$$

$$SSX_1X_2 = \sum_{i=1}^{2}\sum_{j=1}^{2}(Y_{ij} - \overline{Y}_{i+} - \overline{Y}_{i+} - \overline{Y}_{+j} + \overline{Y})^2 \qquad (5.41)$$

The total sum of squares $(SST)$ of the performance response for the general case of $n$ design parameters can be inferred as:

$$SST = SS(1)+SS(2)+,\cdots,SS(n)+SS(1,2)+SS(1,3),\cdots,+SS(1,2,\ldots,n) \quad (5.42)$$

The total sum-of-squares $(SST)$ indicates overall performance variability due to the variance of factors or design parameters. It is also used as a verification criterion to verify the system robust design algorithm. Whereas the partition of the sum-of-squares indicates how much of the total variability $(SST)$ in performance response can be attributed to individual factors and the interaction of factors. In addition, the relative importance of each factor and cross product factor interactions is assessed by dividing the appropriate additive term by the total sum of squares $SST$. For instance, in Table 5.15, $SSX_1/SST$ is the relative influence (sensitivity) of factor $x_1$, and $SSX_1X_2/SST$ is the relative influence (sensitivity) of the interaction of factors $x_1$ and $x_2$. Additive terms with the largest percentages are considered the most important in affecting response variability.

### 5.4.2 Statistical Response Analysis of Microresonators

A microresonator is selected as a representative example of composite microsystems. Use of electrostatic linear comb drive microresonators has been reported in building filters, oscillators [100], and pressure sensors. Electrostatic transduction is commonly used for microresonators because it offers a simplified fabrication process. The circuit-level behavioral models of a electrostatic-comb microresonator is presented in Section 5.1.

The design objectives for the statistical response analysis of the microresonator are as follows:

- Minimize resonant frequency $(\omega_o)$ variability while maintaining on-target response, and
- Maximize motional transconductance $Y_x$

The behavioral model of the microelectromechanical resonator presented in the previous section defines the set of design parameters. Nominal values for the elemental design parameters are listed in Table 5.16. Factor levels are defined using a knowledge of the standard deviation $\sigma$ of the statistical variance distributions due to manufacturing disturbances and are often estimated as $\pm 3\sigma$ [101]. Assumed noise factors with mean and tolerance are listed in Table 5.17.

Hence, a standard three-level orthogonal array $L_{18}$ [92] shown in Table 5.18 with seven columns and three levels can be used to define a partial factorial set of simulations instead of a complete factorial or exhaustive set of simulations. The partial

**Table 5.16** Elemental Design Parameters and Initial Nominal Values

| Parameters | Values | Units |
|---|---|---|
| Folded Flexure Beam Width ($W$) | 2 | $\mu m$ |
| Folded Flexure Beam Length ($L$) | 200 | $\mu m$ |
| Number of Interdigitated Fingers ($N$) | 12 | |
| Thickness of the Fingers ($h$) | 2 | $\mu m$ |
| Gap Between the Fingers ($d$) | 2 | $\mu m$ |
| Young's Modulus ($E$) | 150 | GPa |
| Bias Voltage ($V_P$) | 80 | V |

**Table 5.17** Noise Factors for Microresonator

| | Tolerance Level | | | |
|---|---|---|---|---|
| Noise Factors | (-1) | (0) | (+1) | Unit |
| $W$ | -0.1 | 0 | 0.1 | $\mu m$ |
| $L$ | -1 | 0 | 1 | $\mu m$ |
| $h$ | -0.2 | 0 | 0.2 | $\mu m$ |
| $d$ | -0.2 | 0 | 0.2 | $\mu m$ |
| $E$ | -5 | 0 | 5 | GPa |

factorial set of simulations yield the data required for performance variability reduction and sensitivity analysis.

To measure the effect of performance variability reduction, an initial analysis of variance is conducted for the performance metrics or responses of resonant frequency $\omega_o$ and motional transconductance $Y_x$; the results are given in Tables 5.19 and 5.20. The relatively high values of the sum of squares of the performance response, calculated as

$$sum\ of\ squares = 3(\bar{y}(+1) - \bar{y})^2 + 3(\bar{y}(0) - \bar{y})^2 + 3(\bar{y}(-1) - \bar{y})^2$$

indicates low design robustness.

To determine the settings of the microresonator design parameters (control factors), at which the noise factor influence is minimized, performance variability reduction assumes the levels given in Table 5.21. Three levels are computed for each control factor. Level 0 denotes the nominal (initial) design point settings. Levels -1 and +1 bracket the level 0 settings and define a parametric range sufficient to exercise the nonlinear relationships between factor variances and performance response variances, i.e. resonant frequency and motional transconductance.

The performance variability reduction procedure is illustrated in Figure 5.19. Each

**Table 5.18** Orthogonal Array for Resonant Frequency $\omega_o$ and Motional Transconductance $Y_x$

| No. | $W$ | $L$ | $E$ | $d$ | $h$ | $e$ | $e$ |
|---|---|---|---|---|---|---|---|
| 1 | -1 | -1 | -1 | -1 | -1 | -1 | -1 |
| 2 | 0 | 0 | 0 | 0 | 0 | 0 | -1 |
| 3 | +1 | +1 | +1 | +1 | +1 | +1 | -1 |
| 4 | -1 | -1 | 0 | +1 | 0 | +1 | -1 |
| 5 | 0 | 0 | +1 | -1 | +1 | -1 | -1 |
| 6 | +1 | +1 | -1 | 0 | -1 | 0 | -1 |
| 7 | -1 | 0 | -1 | +1 | +1 | 0 | 0 |
| 8 | 0 | +1 | 0 | -1 | -1 | +1 | 0 |
| 9 | +1 | -1 | +1 | 0 | 0 | -1 | 0 |
| 10 | -1 | +1 | +1 | -1 | 0 | 0 | 0 |
| 11 | 0 | -1 | -1 | 0 | +1 | +1 | 0 |
| 12 | +1 | 0 | 0 | +1 | -1 | -1 | 0 |
| 13 | -1 | 0 | +1 | 0 | -1 | +1 | +1 |
| 14 | 0 | +1 | -1 | +1 | 0 | -1 | +1 |
| 15 | +1 | -1 | 0 | -1 | +1 | 0 | +1 |
| 16 | -1 | +1 | 0 | 0 | +1 | -1 | +1 |
| 17 | 0 | -1 | +1 | +1 | -1 | 0 | +1 |
| 18 | +1 | 0 | -1 | -1 | 0 | +1 | +1 |

**Table 5.19** Initial Analysis of Variance for Resonant Frequency $\omega_o$

| | Level Mean (kHz) | | | Sum of | | |
|---|---|---|---|---|---|---|
| Parameters | lower (-1) | normal (0) | high (+1) | Squares | Percentage | Order |
| $W$ | 15.907 | 17.183 | 18.495 | 10.054 | 32.90% | 1 |
| $L$ | 17.350 | 17.178 | 17.057 | 0.131 | 0.43% | 4 |
| $E$ | 16.895 | 17.188 | 17.502 | 0.553 | 1.81% | 3 |
| $h$ | 16.320 | 17.208 | 18.058 | 4.529 | 14.82% | 2 |
| Total | - | - | - | 30.561 | 100% | - |
| Overall mean | | 17.195 | | - | - | - |

**Table 5.20** Initial Analysis of Variance for Motional Transconductance $Y_x$

| | Level Mean ($10^{-9}s$) | | | Sum of | | |
|---|---|---|---|---|---|---|
| Parameters | lower (-1) | normal (0) | high (+1) | Squares | Percentage | Order |
| $d$ | 3.7253 | 3.0182 | 2.4818 | 2.3340 | 23.92% | 1 |
| $h$ | 2.4655 | 3.0595 | 3.7003 | 2.2883 | 23.45% | 2 |
| $V_p$ | 2.9057 | 3.0487 | 3.2710 | 0.2033 | 2.08% | 3 |
| Total | - | - | - | 9.7575 | 100% | - |
| Overall mean | | 3.0751 | | - | - | - |

**Table 5.21** Control Factors for the Microresonator

| Parameter | Levels | | |
|---|---|---|---|
| | (-1) | (0) | (+1) |
| $W$ | 1 | 2 | 4 |
| $L$ | 100 | 200 | 400 |
| $d$ | 1 | 2 | 4 |
| $h$ | 1 | 2 | 4 |
| $N$ | 6 | 12 | 24 |
| $V_p$ | 40 | 80 | 160 |

row of the control orthogonal array represents a different trial design. For each trial design, the eighteen (18) test conditions are determined and exercised using the circuit microresonator simulation model.

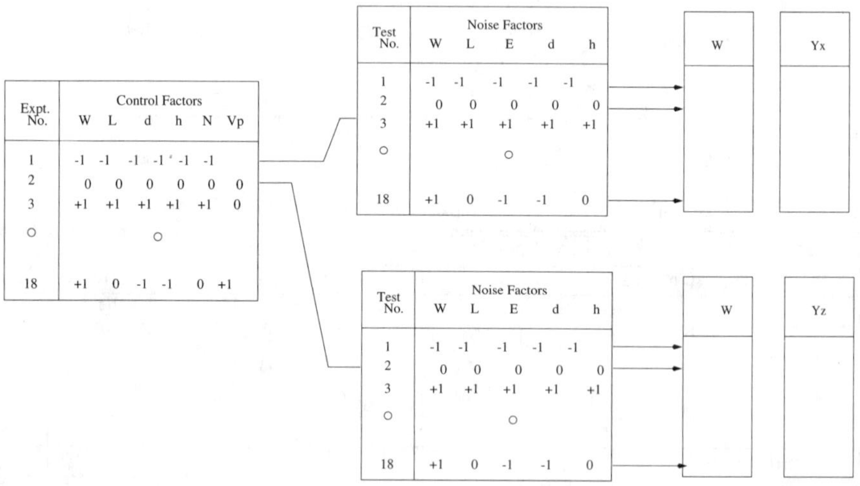


**FIGURE 5.19**
**Experiment design. Each row of the control orthogonal array represents a different trial design.**

Simulation results provide the data for computing the applicable signal-to-noise ratio $SNR$ for the resonant frequency $\omega_o$ and motional transconductance $Y_x$ performance responses. Performance variability reduction results are tabulated in Tables 5.22, 5.23, and 5.24 and plotted in Figures 5.20, 5.21, and 5.22.

Based on the results of the performance variability reduction, the following obser-

**Table 5.22** Analysis of $SNR$ Ratio for Resonant Frequency $\omega_o$

| Parameter | Average $SNR^*$ by level | | |
|---|---|---|---|
| | lower (-1) | normal (0) | high (+1) |
| $W$ | 17.1537 | 21.5687 | 24.5025 |
| $L$ | 20.7956 | 20.8601 | 21.5692 |
| $h$ | 18.7551 | 21.5486 | 22.9212 |
| $N$ | 21.5145 | 20.8627 | 20.8477 |
| Overall mean | $SNR$ =21.0750 | | |

$$*SNR = -10log_{10}\left[\frac{\sigma^2}{\mu^2}\right]$$

**Table 5.23** Analysis of Mean for Resonant Frequency $\omega_o$

| Parameter | Average $\mu$ by level (kHz) | | |
|---|---|---|---|
| | lower (-1) | normal (0) | high (+1) |
| $W$ | 9.153 | 26.614 | 63.180 |
| $L$ | 64.749 | 24.518 | 9.678 |
| $h$ | 26.893 | 38.129 | 33.924 |
| $N$ | 29.556 | 37.249 | 32.141 |
| Overall mean | $\mu$ =32.982 | | |

**Table 5.24** Analysis of $SNR$ Ratio for Motional Transconductance $Y_x$

| Parameter | Average $SNR^*$ by level | | |
|---|---|---|---|
| | lower (-1) | normal (0) | high (+1) |
| $d$ | -39.5983 | -51.1616 | -63.0754 |
| $h$ | -63.9759 | -51.0591 | -38.8003 |
| $N$ | -63.3196 | -51.2784 | 39.2372 |
| $V_p$ | -63.3196 | -51.2784 | -39.2372 |
| Overall mean | $SNR$ =-51.2784 | | |

$$*SNR = -10log_{10}\left[\frac{1}{n}\sum_{i=1}^{n}\frac{1}{y_i^2}\right]$$

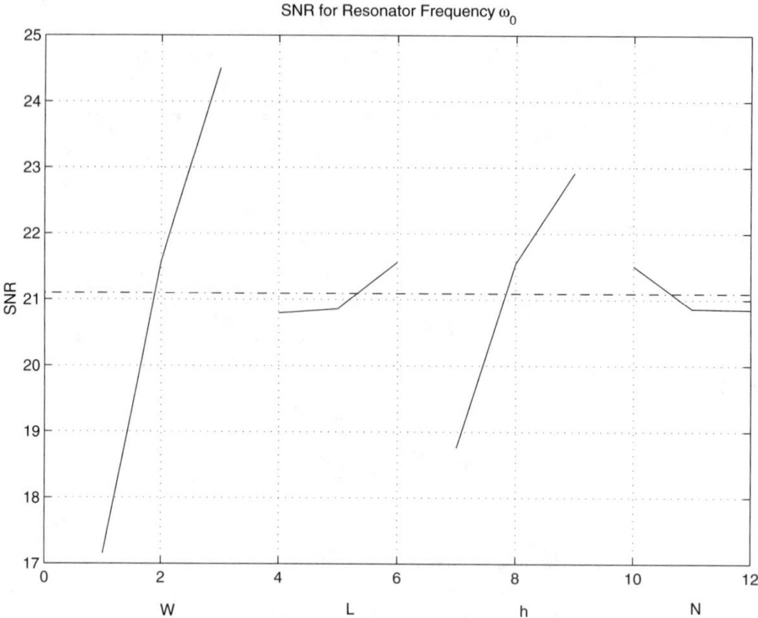


**FIGURE 5.20**
**Plot of control factor effects of SNR on resonant frequency**

vations can be made concerning robust microelectromechanical resonator behavioral modeling and design:

- Folded flexure beam width ($W$) has a significant effect on both $SNR$ ratio and mean of resonant frequency; this result is consistent with the observation that the folded flexure beam is the principal mechanical vibratory structure. The $W_3$ level is the "optimal" design parameter setting as the corresponding $SNR$ is maximal and, conversely, the design performance response variability is minimal.

- Folded flexure beam length ($L$) has little effect on $SNR$ ratio for resonant frequency, but a significant effect on the mean of resonant frequency. Hence, the folded flexure beam length is a good candidate parameter to use to tune the performance resonant frequency target.

- Thickness of the finger ($h$) has a significant effect on both the $SNR$ of resonant frequency and motional transconductance; $h_3$ is the "optimal" parameter setting to maximize both performance metric $SNR$s.

- The gap between the comb drive interdigitated fingers ($d$) has a significant effect on $SNR$ of motional transconductance; $d_1$ is the "optimal" parameter setting.

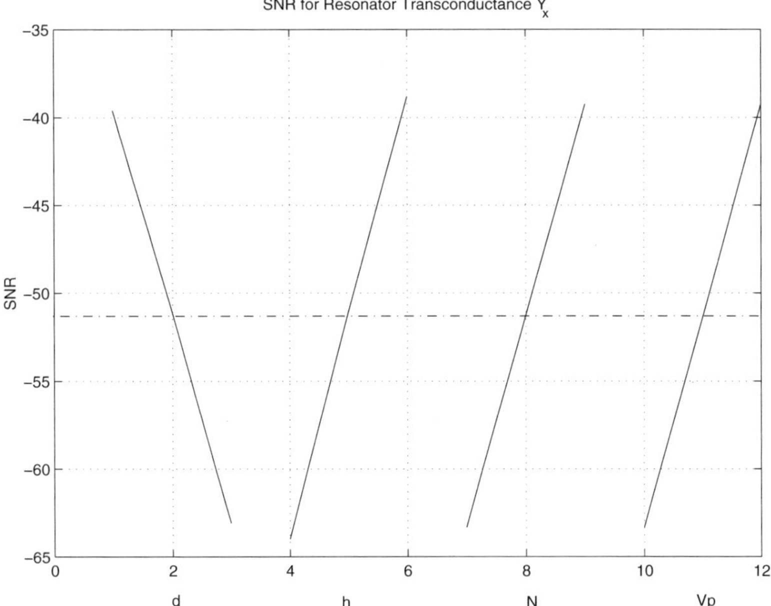


**FIGURE 5.21**
**Plot of control factor effect of mean on resonant frequency**

- The number of fingers in the comb drive ($N$) has considerable effect on $SNR$ of motional transconductance, but little effect on resonant frequency.
- Bias Voltage ($V_P$) has a significant effect on $SNR$ of motional transconductance; $V_{P_3}$ is the "optimal" parameter setting.

Taking into consideration the above tradeoffs and additional issues concerning manufacturability, an optimal design parameter setting is given in Table 5.25. Tables 5.26 and 5.27 show results on conducting the analysis of variance for the optimal design point. Comparing the final analysis of variance for the optimal design point with the initial analysis of variance given in Tables 5.19 and 5.20 reveals an improvement in both design objectives. The less $SST$ in Table 5.26, 12.282, comparing to that in Table 5.19, 30.561, verifies the correctness of the system robust design algorithm. The motional transconductance has been significantly increased, as shown by the higher means values. Also, the variability of the resonant frequency has been significantly decreased, as shown by the decrease in the total sum of squares, indicating improved design robustness. The partitions of the sum of squares of the final analysis of variance defines the relative contributions of the design parameter variance to overall design performance variance. This data is a measure of design parameter sensitivity and is graphically shown in Figures 5.23 and 5.24.

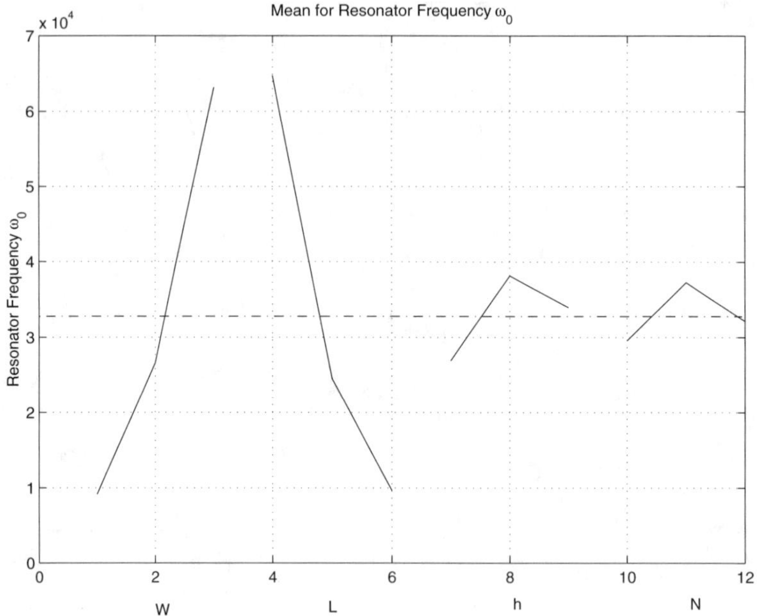


**FIGURE 5.22**
**Plot of control factor effect of SNR on motional transconductance $Y_x$**

**Table 5.25** Optimal Design

| $W$ | $L$ | $E$ | $d$ | $h$ | $N$ | $V_p$ | Frequency (kHz) | Transconductance ($10^{-9}s$) |
|---|---|---|---|---|---|---|---|---|
| 3 | 378 | 150 | 1 | 4 | 20 | 80 | 17.195 | 133 |

**Table 5.26** Analysis of Variance for Noise Factors on Resonant Frequency $\omega_o$ Based on Optimal Design

| Parameter | Level Mean (kHz) | | | Sum of | | |
|---|---|---|---|---|---|---|
| | lower(-1) | Normal(0) | high(+1) | Squares | Percentage | Order |
| $W$ | 16.338 | 17.192 | 18.063 | 4.4595 | 36.31% | 1 |
| $L$ | 17.277 | 17.190 | 17.126 | 0.0343 | 0.28% | 4 |
| $E$ | 16.906 | 17.196 | 17.491 | 0.5133 | 4.18% | 3 |
| $h$ | 16.762 | 17.200 | 17.631 | 1.1318 | 9.22% | 2 |
| Total | - | - | - | 12.282 | | |
| Overall mean | | 17.195 | | | | |

**Table 5.27** Analysis of Variance for Noise Factors on Motional Transconductance $Y_x$ Based on Optimal Design

| Parameters | Level Mean ($10^{-9}s$) | | | Sum of Squares | Percentage | Order |
|---|---|---|---|---|---|---|
| | lower(-1) | Normal(0) | high(+1) | | | |
| $d$ | 208.3625 | 133.3545 | 92.3757 | 20758 | 45.70% | 1 |
| $h$ | 129.4642 | 145.0317 | 159.5968 | 1362 | 3% | 2 |
| $V_p$ | 136.8362 | 143.3260 | 153.9305 | 447 | 0.98% | 3 |
| Total | - | - | - | 45419 | 100% | - |
| Overall mean | 144.6976 | | | - | - | - |

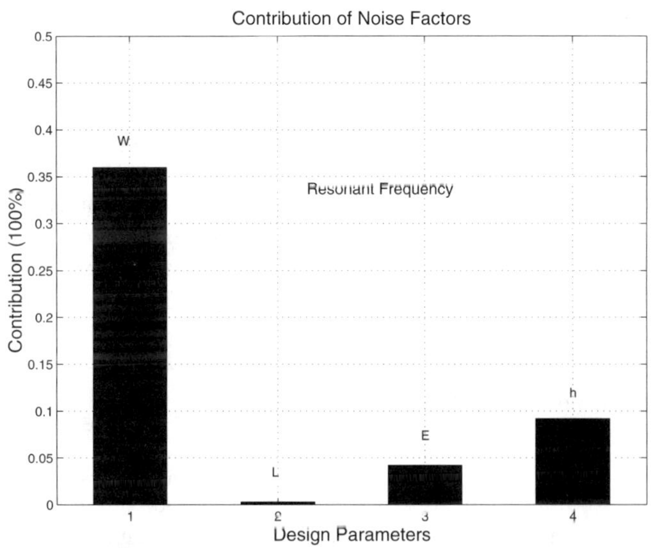


**FIGURE 5.23**
**Parameter contributions to resonant frequency variance. The resonant frequency is most sensitive to the variance of the $w$.**

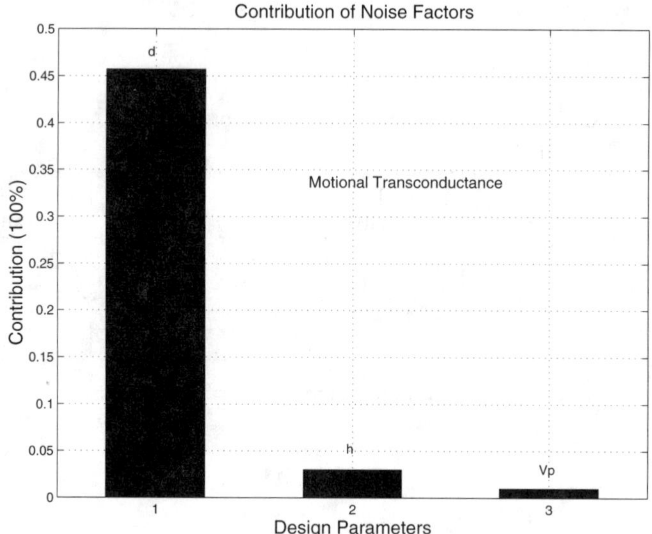


**FIGURE 5.24**
**Parameter contributions to transconductance variance. The transconductance sensitivity to the variation of $d$ is maximum.**

This section addresses a robust design methodology for composite microsystem design based on Taguchi robust parameter design and statistical process control methods. Behavioral models are used to conduct sample simulations. The resulting data is analyzed and used to determine model parameter settings that minimize the variability in overall performance response to fluctuations in design parameters. The relative contributions of the individual model parameters to the minimized design variability is also determined. In addition, the correctness of the system robust design algorithm has been verified using the statistical total variability ($SST$).

### 5.4.3 Design Optimization of Microvalves

Sum-of-squared values of performance response decomposes the total variability of response into separate elements representing the variance of each factor and their interaction, and could present the useful performance variability information [97], but the variance analyses could not directly represent the geometric relationships between the system performance and the design parameters. Statistical response analyses can be studied using multivariate linear regression modeling. This study helps the designer to understand the casual relationships between how fluctuations in design parameters shift the design point and associated system behavior, and indicates which factors need to be more carefully analyzed during design and more carefully controlled during manufacturing and operating [31]. This study can help to solve the optimal design tradeoff to find the appropriate design solution, and verify the efficiency of the statistical analysis.

Depending on the system modeling of the microvalve, the design parameters which influence performance responses, and their nominal design values are shown on Table 5.28.

**Table 5.28** Elemental Parameters and Initial Nominal Design Values

| Parameters | Values | Units |
|---|---|---|
| Length of the cantilever ($L$) | 1600 | $\mu m$ |
| Width of the cantilever($b'$) | 1000 | $\mu m$ |
| Thickness of the cantilever ($h'$) | 15 | $\mu m$ |
| Height of the valve seat ($h$) | 50 | $\mu m$ |
| Length of the valve seat ($l_1$) | 5 | $\mu m$ |
| Width of the valve seat ($b$) | 400 | $\mu m$ |
| Length of the cantilever over valve seat($l_2$) | 100 | $\mu m$ |
| Young's Modulus ($E$) | 146.9 | GPa |
| Air Pressure($P_a$) | 100000 | $P_a$ |

The design objective of microvalve performance is to minimize the variation of the static flow rate $\Phi$ due to the fluctuation of design parameters, while maintaining the on-target performance. Note that since $L, b', h', l_1, b, h, l_2, p$ are subject to manufacturing variations, the values of each variable is bounded by $\pm 3\sigma$, where $\sigma$ is the standard deviation for the parameter. The resulting noise factors with mean and tolerance are listed in Table 5.29.

**Table 5.29** Noise Factors for Microvalve

| Noise Factors | Tolerance Level | | | Unit |
|---|---|---|---|---|
| | (-1) | (0) | (+1) | |
| $L$ | -0.2 | 0 | 0.2 | $\mu m$ |
| $b'$ | -0.2 | 0 | 0.2 | $\mu m$ |
| $h'$ | -0.2 | 0 | 0.2 | $\mu m$ |
| $l_1$ | -0.2 | 0 | 0.2 | $\mu m$ |
| $b$ | -0.2 | 0 | 0.2 | $\mu m$ |
| $h$ | -0.2 | 0 | 0.2 | $\mu m$ |
| $l_2$ | -0.2 | 0 | 0.2 | $\mu m$ |
| $p$ | -1 | 0 | 1 | Pa |

The Young's modulus ($E$) is related to the design material, and is assumed to maintain constant. Moreover, in order to reduce the sensitivity to manufacturing variations of the structural etch, some of the design parameters are correlated, for instance the design parameters of the vantilever, hence their variance is the same.

To determine the design parameter setting for the microvalve robust design, at which the noise factor influence is minimized, depending on the microvalve design envelope and its characteristic resonance frequency, which is related with the microvalve geometric value, three design levels are set shown in Table 5.30, the initial normal design points are defined at level 0. Parametric wider range is defined to identify the nonlinearity of the relationship between the control factors and performance response, and to avoid the operating microvalve near the resonance frequency, which could make the flow rate unstable.

**Table 5.30** Control Factors for the Microvalve

| Parameter | Levels | | |
|---|---|---|---|
| | (-1) | (0) | (+1) |
| $L$ | 1280 | 1600 | 1920 |
| $b'$ | 800 | 1000 | 1200 |
| $h'$ | 12 | 15 | 18 |
| $h$ | 40 | 50 | 60 |
| $l_1$ | 4 | 5 | 6 |
| $b$ | 320 | 400 | 480 |
| $l_2$ | 80 | 100 | 120 |

Therefore, a standard orthogonal array $L_{18}$ with seven columns and three factor levels defined as fractional factorial design instead of a complete factorial simulation can be used to derive the inner and outer array simulation.

At the normal design point, (5.43) shows the multivariate linear regression model of the microvalve static flow rate $\Phi(ml/min)$, obtained using the statistical response analysis techniques.

$$\Phi = -15083.42+2.407\times L+2.407\times b'+2.407\times h'+0.065\times h+14.654\times b+0.088\times p \tag{5.43}$$

The coefficient of determination $R^2$ and the adjusted coefficient of determination $R^2_{adj}$ statistic show an very good linear fit.

$$R^2 = 1.00 \quad R^2_{adj} = 1.00 \tag{5.44}$$

**Table 5.31** Array Simulation Design for the Static Flow Rate $\Phi$

| No.(Outer Array) | $L$ | $b'$ | $h'$ | $h$ | $l_1$ | $b$ | $l_2$ |
|---|---|---|---|---|---|---|---|
| 1 | -1 | -1 | -1 | -1 | -1 | -1 | -1 |
| 2 | 0 | 0 | 0 | 0 | 0 | 0 | -1 |
| 3 | +1 | +1 | +1 | +1 | +1 | +1 | -1 |
| 4 | -1 | -1 | 0 | +1 | 0 | +1 | -1 |
| 5 | 0 | 0 | +1 | -1 | +1 | -1 | -1 |
| 6 | +1 | +1 | -1 | 0 | -1 | 0 | -1 |
| 7 | -1 | 0 | -1 | +1 | +1 | 0 | 0 |
| 8 | 0 | +1 | 0 | -1 | -1 | +1 | 0 |
| 9 | +1 | -1 | +1 | 0 | 0 | -1 | 0 |
| 10 | -1 | +1 | +1 | -1 | 0 | 0 | 0 |
| 11 | 0 | -1 | -1 | 0 | +1 | +1 | 0 |
| 12 | +1 | 0 | 0 | +1 | -1 | -1 | 0 |
| 13 | -1 | 0 | +1 | 0 | -1 | +1 | +1 |
| 14 | 0 | +1 | -1 | +1 | 0 | -1 | +1 |
| 15 | +1 | -1 | 0 | -1 | +1 | 0 | +1 |
| 16 | -1 | +1 | 0 | 0 | +1 | -1 | +1 |
| 17 | 0 | -1 | +1 | +1 | -1 | 0 | +1 |
| 18 | +1 | 0 | -1 | -1 | 0 | +1 | +1 |
| No.(Inner Array) | $(L, b', h')$ | $h$ | $l_1$ | $b$ | $l_2$ | $p$ | $e$ |

Using the regression model on the (5.43), Table 5.32 shows the analysis of variance for noise factors on the static flow rate $\Phi$. The main effect, $\partial\Phi/\partial x_i$, presents the average variation of the performance response due to the unit fluctuation of the individual design parameter. Total sensitivity shows the system performance robust. Because of the correlation between design parameters, $L, b, h'$, their order-rank is the same.

**Table 5.32** Analysis of Variance for Noise Factors on Flow Rate $\Phi$

| Parameters | $L$ | $b'$ | $h'$ | $h$ | $b$ | $p$ |
|---|---|---|---|---|---|---|
| $\partial\Phi/\partial x_i$ | 2.407 | 2.407 | 2.407 | 0.065 | 14.654 | 0.088 |
| Order | 2 | 2 | 2 | 6 | 1 | 5 |
| Unit Robust | 22.028 | | | | | |
| $SST$ | 2.8485e-06 | | | | | |

** The effects of $l_1, l_2$ are too small and can be ignored.

With the VHDL-AMS modeling, simulation results provide the data to compute the applicable signal-to-noise ratio $SNR$ for the static flow rate. Table 5.33, 5.34, and Figure 5.25, 5.26 show the control factor effect of SNR and the performance mean value, and some observations can be made concerning the robust system design.

**Table 5.33** Analysis of $SNR$ Ratio for the Static Flow Rate $\Phi$

| Parameter | Average $SNR^*$ by level | | |
|---|---|---|---|
| | lower (-1) | normal (0) | high (+1) |
| $L$ | 66.55 | 66.02 | 64.75 |
| $b'$ | 65.34 | 65.64 | 66.34 |
| $h'$ | 64.18 | 66.24 | 66.91 |
| $h$ | 65.36 | 65.67 | 66.30 |
| $l_1$ | 66.04 | 65.37 | 65.92 |
| $b$ | 64.75 | 65.81 | 66.76 |
| $l_2$ | 65.97 | 65.89 | 65.47 |

$*SNR = -10log_{10}\left[\frac{\sigma^2}{\mu^2}\right]$

**Table 5.34** Analysis of Mean for the Static Flow Rate $\Phi$

| Parameter | Average $\mu$ by level (ml/min) | | |
|---|---|---|---|
| | lower (-1) | normal (0) | high (+1) |
| $L$ | 4644.37 | 5818.98 | 7116.86 |
| $b'$ | 4657.49 | 5811.72 | 7111.01 |
| $h'$ | 5964.13 | 5756.31 | 5859.78 |
| $h$ | 5955.16 | 5749.83 | 5875.23 |
| $l_1$ | 5962.17 | 5747.58 | 5870.46 |
| $b$ | 4654.85 | 5811.43 | 7113.93 |
| $l_2$ | 6097.36 | 5743.05 | 5739.81 |

- Cantilever length ($L$) has a significant effect on both $SNR$ ratio and the mean of flow rate $\Phi$. The $L_1$ level is the "optimal" design parameter setting as the corresponding $SNR$ is maximal and, conversely, the design performance response variability is minimal.

- Cantilever width ($b'$) also has a significant effect on $SNR$ ratio, but unlike cantilever length ($L$), it has the opposite effect on the mean of flow rate $\Phi$, hence, it can be used as a good candidate parameter to tune the system performance target.

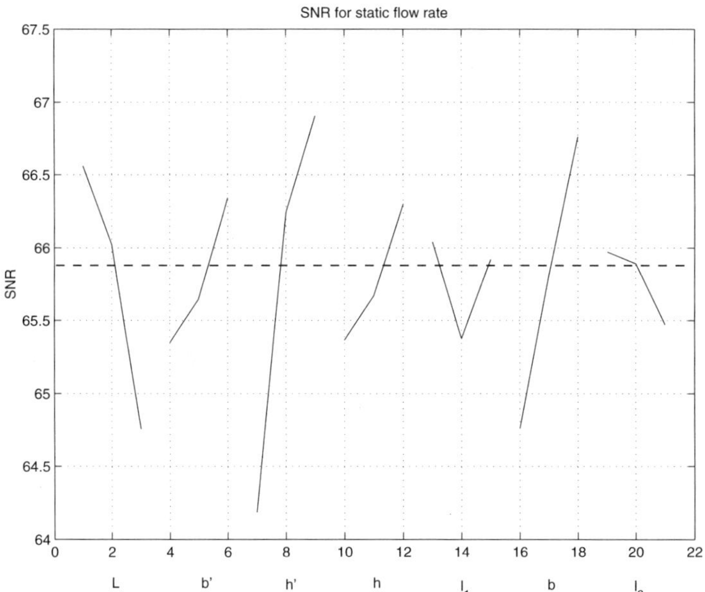


**FIGURE 5.25**
**Plot of control factor effect of SNR on flow rate**

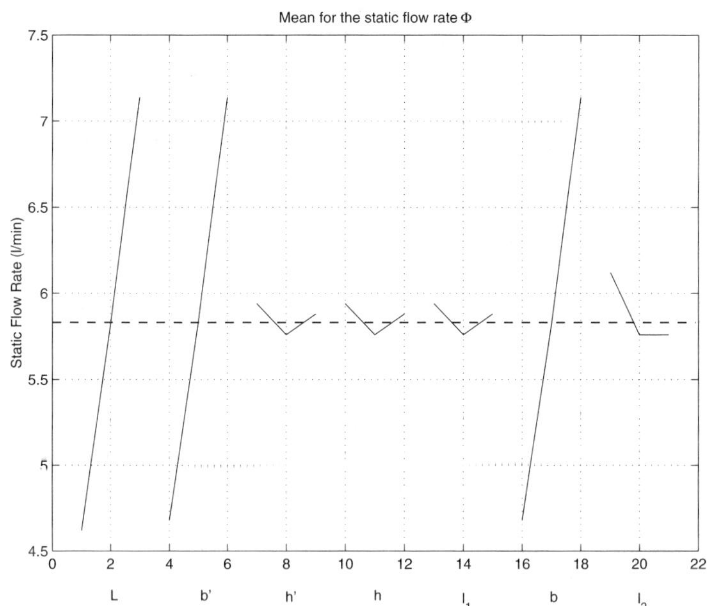


**FIGURE 5.26**
**Plot of control factor effect of mean on flow rate**

- Cantilever thickness ($h'$) has a huge effect on $SNR$ ratio but little effect on the mean of flow rate $\Phi$, $h'_3$ is the "optimal" parameter setting to maximize the $SNR$.
- Height of the valve seat ($h$) has a significant effect on $SNR$ ratio but little effect on the mean of flow rate $\Phi$, $h_3$ is the "optimal" parameter setting to maximize the $SNR$.
- Length of the valve seat ($l_1$) has two design levels, $l_1$ or $l_3$, possessing the nearly same $SNR$, and having little effect on the mean of flow rate $\Phi$, it also can be used as a candidate parameter to tune system performance.
- Width of the valve seat ($b$) has a significant effect on both $SNR$ ratio and the mean of flow rate $\Phi$, the $b_3$ level is the "optimal" design parameter setting to maximize the corresponding $SNR$.
- Length of the cantilever over valve seat($l_2$) has considerable effect on $SNR$ ratio and little effect on flow rate.

Taking into consideration the above observations, and additional issues concerning the manufacturability, a comparison of parameter settings for the nominal design point and a potential optimal design is given in Table 5.35. Both designs possess the same static flow rate, $5861.58(ml/min)$.

**Table 5.35** Optimal Design

| | $L$ | $b'$ | $h'$ | $h$ | $l_1$ | $b$ | $l_2$ |
|---|---|---|---|---|---|---|---|
| Nominal Design Point | 1600 | 1000 | 15 | 50 | 5 | 400 | 100 |
| Optimal Design Point | 1280 | 1145 | 18 | 40 | 4 | 440 | 80 |

With the coefficient of determination $R^2$ and the adjusted coefficient of determination $R^2_{adj}$ statistic being equal to 1, results show the very good linear fit between the microvalve static flow rate $\Phi(ml/min)$ and the design point at the optimal design point.

$$\Phi = -15083.42 - 1.200 \times L - 1.200 \times b' - 1.200 \times h' + 0.072 \times h$$
$$-0.002 \times l_1 + 13.321 \times b - 0.002 \times l_2 + 0.440 \times p \tag{5.45}$$

Using the regression model on the (5.45), Table 5.36 shows the analysis of variance for noise factors on the static flow rate $\Phi$.

Comparing the final analyses in Table 5.32 and Table 5.36, the decrease of the main effect, $\partial\Phi/\partial x_i$, the unit robust and the total sum of square of static flow rate show

**Table 5.36** Analysis of Variance for Noise Factors on Flow Rate $\Phi$

| Parameters | $L$ | $b'$ | $h'$ | $h$ | $l_1$ | $b$ | $l_2$ | $p$ |
|---|---|---|---|---|---|---|---|---|
| $\partial\Phi/\partial x_i$ | 1.200 | 1.200 | 1.200 | 0.072 | 0.002 | 13.321 | 0.002 | 0.440 |
| Order | 2 | 2 | 2 | 6 | 7 | 1 | 7 | 5 |
| Unit Robust | 17.085 | | | | | | | |
| $SST$ | 2.0332e-06 | | | | | | | |

that the system stability has been significantly enhanced. The comparison of main effects of individual design parameters, which measures the disturbing influence of each design parameter, and the comparison of unit robust and $SST$ are graphically shown in Figure 5.27 and 5.28.

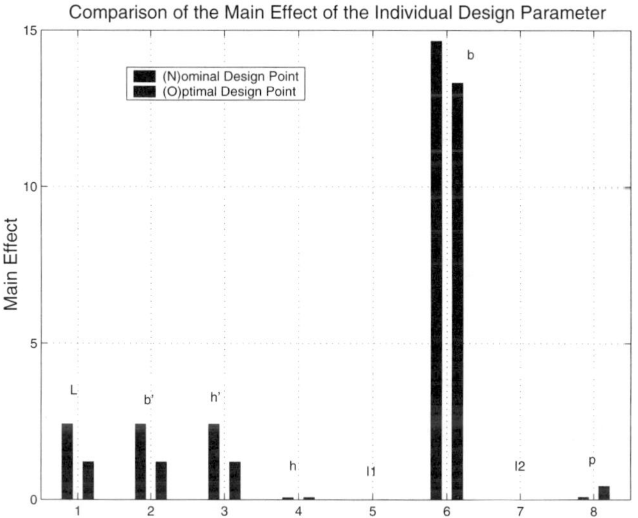


**FIGURE 5.27**
**Comparison of the main effect of the individual design parameter**

By the Taguchi robust parameter design method and sensitivity analysis approach, this section presents a novel application approach of experimental method in the statistical simulation area, which decreases the period of design, increases feed-back speed of the enterprise, and is a novel application of advanced concurrent engineering on MEFS. Meanwhile, (5.43) and (5.45) reasonably represent the complex physical models with the multivariate linear regression models, which can be used to conduct efficient sensitivity information to understand the manufacturing process and operational limits, then suggest the optimal control strategy for further workshop fabrication and operating environment, and it also will benefit new generation de-

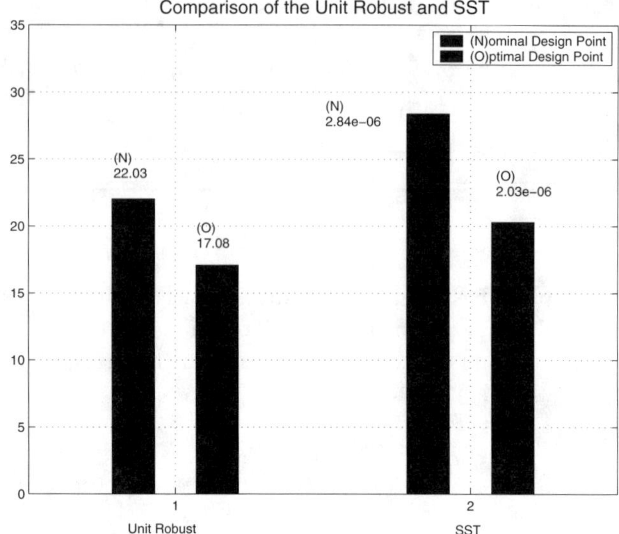


**FIGURE 5.28**
**Comparison of the unit robust and SST**

sign.

## 5.5 Application Flexibility Optimization[2]

A number of design methodologies for microsystems have recently been proposed [97], [98], [102]. These methods lead to robust microsystems that meet performance goals but are relatively insensitive to design parameter variations. However, they are tailored towards "custom microsystems" whose components are designed to operate within a narrow range of system performance. This leads to expensive and inflexible systems that are not amenable to large-volume production. This is one of the main obstacles facing composite microsystems development [99].

We propose a reconfigurable composite microsystem design methodology that leverages hardware/software co-design principles to achieve functional unit reusability.

[2]Reprinted from Proc. 4th Int. Conf. Modeling and Simulation of Microsystems (MSM2001). T. Zhang, K. Chakrabarty and R. B. Fair, Design of Reconfigurable Composite Microsystems based on Hardware/Software Co-design Principles pp. 148-152, 2001. © 2001, with permission from Applied Computational Research Society (ACRS).

The hardware/software co-design method provides design flexibility by allowing software to be compiled efficiently for a modular hardware platform [32]. By focusing on the specific characteristics of composite microsystems, an analogy is developed between microsystem components and the hardware and software components of a microelectronic system.

By partitioning the design parameters of microsystems into two categories: nonreconfigurable "hardware" and reconfigurable "software" design parameters (referred to as $NRDP$s and $RDP$s, respectively), we can make the microsystem performance meet the flexibility requirement and be suitable for a wider range of applications. While the values of the $NRDP$s are determined at fabrication time, the values of the $RDP$s are configured (programmed) during operation. This allows the system to conform to a wider range of performance specifications. The key challenges here include the following: (i) partition the set of design parameters to $NRDP$s and $RDP$s; (ii) determine the values of the $NRDP$s to make the system performance less sensitive to variation of $NRDP$s; (iii) exploit the synergy between $NRDP$s and $RDP$s to increase performance flexibility. Thus, given a range of values that the $RDP$s can take, the goal here is to determine the values of the $NRDP$s such that the range of system performance is maximized under the constraint that the performance is relatively insensitive to the variation of the $NRDP$s. Table 5.37 illustrates the partitioning example for a generic microelectrofluidic system (MEFS).

**Table 5.37** Design Parameters for a Microfluidic System.

| **Abstraction Level** | **Nonreconfigurable Design Parameters** *(Hardware)* | **Reconfigurable Design Parameters** *(Software)* |
|---|---|---|
| System Level | Number of I/O Interface Ports<br>Number of Delivery Channel Buses | Plug in/out Strategy<br>Process Scheduling |
| Component Level | Beam Dimension | Pressure |
| | Channels Diameter | Electrical Voltage |
| | Fabrication Materials | Operating Frequency |

The Taguchi experimental design method [30] provides an efficient method for performance variability reduction, and it is often used for off-line parametric optimization control and high performance design. The basic idea of this method is to identify the parameters or factors most influential in determining a performance metric and to compute an appropriate setting of the parameters. This is done using orthogonal arrays and design of experiments. We use the Taguchi method to ensure that the system performance lies within an acceptable range, and the influence of parametric variations on the system performance is minimized. Statistical response surface analysis studies the system performance variability within a region. Thus, it characterizes the relationships between the basic electrical/mechanical parameters and system performance. This allows a designer to understand how fluctuations in design parameters shift the design point and the associated system behavior, and then to explore the maximal system performance range.

Therefore, the procedure of the application flexibility design optimization is that, at first, on the analogy of the hardware/software co-design principles, we partition the set of design parameters into $NRDP$s and $RDP$s for application flexibility. Then, we use the Taguchi experimental design method to determine the values of $NRDP$s that make the system performance robust, that is, less sensitive to variation of $NRDP$s. Next, we increase the application flexibility of composite microsystems using the response surface methodology. Finally, given a range of values that $RDP$s can take, we design the system such that the range of system performance is maximized under the constraint that the performance is relatively insensitive to the variation of $NRDP$s.

The organization of this section is as follows. The general problem statement and design approach are presented in Section 5.5.1. The Taguchi experiment design method [30], which is described in Section 5.3, is used to determine the value of the $NRDP$s for robust design. We also present the response surface methodology [31], which is used to maximize the performance range for a given programmable set of $RDP$s. Section 5.5.3 further describes the design procedure for achieving application flexibility. Section 5.5.4 presents a case study based on the microvalve, which serves as a case study for an electrostatic microfluidic device.

## 5.5.1 Design Approach

### 5.5.1.1 *RDP*s and *NRDP*s Partitioning

The overall microsystem cost and performance are affected by the partitioning of the design parameters into $NRDP$s and $RDP$s. Based on the characteristics of composite microsystems, the $NRDP$s/$RDP$s partitioning algorithm can be induced into two categories

- *Explicit Partitioning*
  It means that the partitioning of design parameters is explicit, such as the geometric design parameters, which must be determined at the fabrication time.

- *Implicit Partitioning*
  It means that the design parameters can be categorized into $NRDP$s or $RDP$s, the final partitioning is based on the different design objectives.

The partitioning decision depends on the relationship between design parameters, system reliability and cost. Some of the issues influencing the partitioning decision are as follows:

1. *Correlation*
   Correlated parameters must be placed in the same category. These correlations and dependencies are generally determined by the designer. Alternatively, a design rule can be used to automatically extract these correlations. For example, there is significant correlation between the beam width and the perimeter of the moving electrode in accelerometers [103]. Therefore, these two parameters (beam width and perimeter of the electrode) must be placed in the same category.

2. *Ease of control*
   Some design parameters, e.g. fluid pressure and electrical voltage, are relatively easy to control during operation. Therefore, these can be placed in the $RDP$ set to increase application flexibility.

3. *Cost*
   The cost of reconfiguration can also be a driving factor. For example, the channel length in a microvalve is expensive to alter after fabrication. Hence it is preferably placed in the $NRDP$ set to reduce cost.

Depending on the particular feature of composite microsystems, there exist the nonconfigurable design parameters, e.g. mechanical design parameters. Therefore, at the initial state of partitioning, all design parameters can be a *complete nonconfigurable design parameter set*, then with the partitioning criterion, reconfigurable design parameter set can be abstracted from the complete nonreconfigurable design parameter set.

$$
\begin{aligned}
&Initial: \\
&\qquad NRDP_{(init)} = U \\
&\qquad RDP_{(init)} = \emptyset \\
&Partitioning: \\
&\qquad \Downarrow \\
&Final: \\
&\qquad NRDP_{(final)} \bigcup RDP_{(final)} = U \\
&\qquad NRDP_{(final)} \bigcap RDP_{(final)} = \emptyset
\end{aligned}
$$

Where

- $U$ : the complete set of design parameters.
- $NRDP_{(init)}$ : the initial $NRDP$ set.
- $RDP_{(init)}$ : the initial $RDP$ set.
- $NRDP_{(final)}$ : the final $NRDP$ set.
- $RDP_{(final)}$ : the final $RDP$ set.

The $NRDP$ values are determined at manufacturing time, and this provides a nonreconfigurable "hardware" platform. The $RDP$s constitute the "re-programmable software" that run on this platform. In this way, composite microsystems provide design flexibility for product evolution and different application purposes.

#### 5.5.1.2 Nonreconfigurable Platform Design Robustness

One of the system optimization objectives is to find an optimal setting of the $NRDP$s that makes the system performance less sensitive to the fluctuation of the manufacturing process and the operating environment. The Taguchi parameter design method, which is widely used for off-line parametric optimization control and high reliability design [30], is used here to achieve this objective.

By choosing levels of the design parameters, we can make the system performance less sensitive to variation of the set of noise factors. Performance variability is computed using a statistical metric signal-to-noise ratio ($SNR$). System performance is considered a "signal" and parametric fluctuations are considered "noise". Our performance variability reduction method computes the effect of each setting of design parameters on $SNR$ and uses the comparison results to determine the best combination of parameter settings to optimize performance stability [97]. The detail Taguchi experiment design method is discussed in Section 5.4.1.

#### 5.5.1.3 Degree of "Programmability" of a Reconfigurable Design Parameter

The next design objective is to determine the degree of "programmability" of the $RDP$s. Therefore, when $RDP$s run on the robust nonreconfigurable "hardware" platform, a composite microsystem can provide the design flexibility for product evolution and different application purpose. The following factors must be considered in this context:

- *Microsystem energy requirement*
  The energy supply availability for composite microsystems is limited due to miniaturization and integration. Hence, the energy requirement of the $RDP$s must conform to this restriction. For example, the adjustable range of the electrical voltage must lie within the available voltage range.
- *Limitation of physical implementation*
  The limitation of physical implementation is also a key for "programmability" of $RDP$s. For example, the operating frequency of a micropump chamber may be limited by the feasibility of physical implementation.
- *Fabrication technology and integration level*
  As composite microsystems become smaller, the fabrication technology and integration levels also limit the range of $RDPs$.

- *Operational reliability*
  Higher degree of "programmability" of $RDP$s may lead to operational reliability problems, hence it may be more difficult to maintain the accurate control at a wider range.

Therefore, designers should consider the above constraints to determine the degree of "programmability" for the $RDP$s.

### 5.5.2 Determining the Performance Flexibility

Based on the certain setting of $NRDP$s and the determined programmability of $RDP$s, the composite microsystem performance flexibility can be obtained. The performance range is from the lowest performance to highest performance, and the response surface methodology can be used to identify this performance flexibility.

The response surface methodology can be used to directly represent the geometric relationships between the system performance and design parameters. This helps the designer to understand the causal relationships between how design parameters shift the design point and associated system behavior. Since the adjustable variation scope of $RDP$s is usually limited, we assume that the relationship between $RDP$s and system response can be represented as a unimodal function. This implies that on the system response surface, there is exactly one point possessing the minimum performance value and exactly one point possessing the maximum performance value, as shown in Figure 5.29. Therefore, the local optimal design point is also the global optimal design point in this design space. While we make the unimodal function assumption here to illustrate our approach, we can handle a system with multimodal response surfaces through piecewise approximation techniques. In this case, as well as for the case where the system performance is not a continuous function of the $RDP$ setting, we can use iterative search over subintervals, in which the performance is a unimodal and continuous function, then we compare the performance upon each subinterval to get the optimal solution.

The minimum and maximum performance points can be formed via iterative search algorithms. When there is just one $RDP$, the relationship between system performance and the $RDP$ can be represented with a curve in the X-Y plane, and a one-dimensional iterative search method, such as *Golden section*, or *Fibonacci* search method can be used to find the minimum and maximum performance points [101]. If the number of $RDP$s is greater than one, the response surface can be used to represent the relationship between system performance and $RDP$s. An iterative gradient search method, such as *Steepest ascent/descent* [31], can be used to find the optimal points.

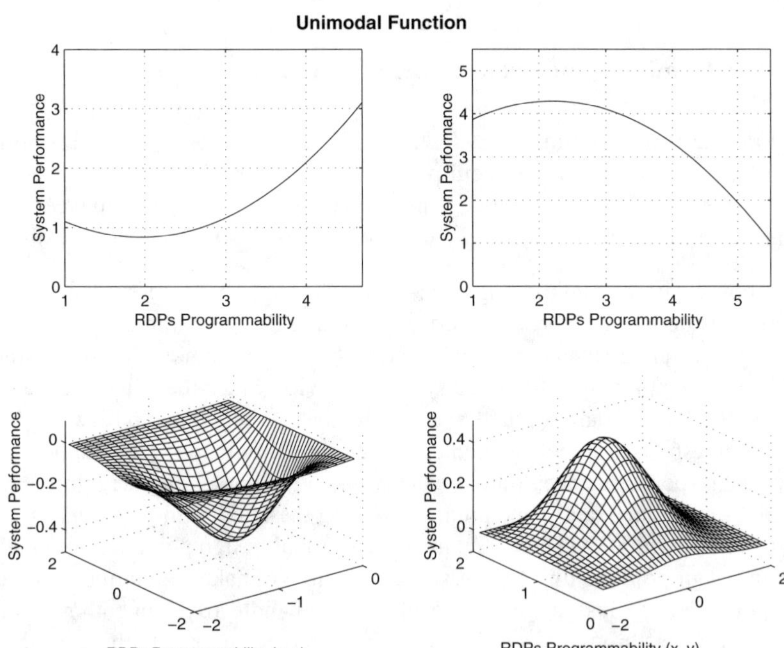


**FIGURE 5.29**
**Unimodal function for two and three dimensions.**

### 5.5.3 Optimization Procedure

The goal of $NRDP$s/$RDP$s co-design is to obtain wider system performance within the feasible programmability range of $RDP$s and a robust setting of the nonreconfigurable "hardware" platform. Since this optimization problem involves multiple objectives, designers need to trade off each objective to get an appropriate design result. Therefore, the proposed optimization procedure includes six steps:

1. Depending on the partitioning criterion, the design parameters are grouped into $RDP$ and $NRDP$ sets.
2. Select a series of settings of $NRDP$s as a "hardware" platform.
3. Determine the degree of programmability of $RDP$s.
4. Using the response surface method and an iterative search algorithm, the minimum and maximum system performance values and related $RDP$ values are found within the determined programmability of $RDP$s.
5. The system robustness (insensitivity to the variation of $NRDP$s) is represented using $SNR$, and optimized using the Taguchi robust design methodology. Since the $SNR$ value for a certain setting of $NRDP$s also depends on the setting of $RDP$s within their programmability range, the $SNR$ for this setting of $NRDP$s may vary with the individual value of $RDP$s. However, with the unimodal assumption, it is reasonable to estimate the robustness for a certain setting of $NRDP$s using the average of $SNR$s which are calculated at the $RDP$ nominal setting value, and the $RDP$ values corresponding to the minimum and maximum performance values. The average of $SNR$ value, $\overline{SNR^i}$, for the $i^{th}$ setting of the $NRDPs$ is given by

$$\overline{SNR^i} = \frac{SNR^i_\alpha + SNR^i_\beta + SNR^i_\gamma}{3}, i = 1, 2, ..., n \tag{5.46}$$

   where:

   - $SNR^i_\alpha$ : $SNR$ value for the $i^{th}$ setting of the $NRDP$s with the $RDP$ setting at the minimum performance value.
   - $SNR^i_\beta$ : $SNR$ value with the nominal setting of $RDP$s on the $i$th $NRDP$ setting.
   - $SNR^i_\gamma$ : $SNR$ value with the $RDP$ setting possessing the maximum performance value on the $i$th $NRDP$ setting.
   - $n$ is the total number of the setting of $NRDP$s.

6. Calculate the main effect for each design parameter at a particular setting. Based on these main effect values, we can obtain the desired performance flexibility and robustness.

### 5.5.4 Case Study: Microvalve Modeling and Optimal Design

In this section, a case study is presented for an electrostatic microfluidic device, the microvalve. Based on the physical principles of operation of the microvalve and the verified circuit-level model discussed in Section 4.4, a microvalve behavior model is developed using SystemC. The final optimized design result ensures robustness and a wider performance range for application flexibility.

#### 5.5.4.1 Optimal Design

Depending on the physical principles of the microvalve and the partitioning criteria, the design parameters can be grouped into the $NRDP$s and $RDP$s sets. The geometric design parameters are grouped into the set of $NRDP$s, and the pressure difference $p$ is placed in the $RDP$ set. The partitioning is shown in Table 5.38, where $L$ is the length of the cantilever, $b'$ is the width of the cantilever, $h'$ is the thickness of the cantilever, $h$ is the height of the valve seat, $l_1$ is the length of the valve seat, $b$ is the width of the valve seat, $l_2$ is the length of the cantilever over valve seat, and $E$ is Young's Modulus.

**Table 5.38** $NRDP$s and $RDP$s.

| $NRDP$s | $RDP$s |
|---|---|
| $L, b', h', h, l_1, b, l_2, E$ | $p$ |

Our design objective is to minimize the variation of the overall flow rate $\Phi$ due to the fluctuation of design parameters. Here, we assume that the design parameter tolerances are $\pm 0.2\mu m$.

**Table 5.39** Tolerance for NRDPs.

| $NRDP$s | $L$ | $b'$ | $h'$ | $h$ | $l_1$ | $b$ | $l_2$ |
|---|---|---|---|---|---|---|---|
| Tolerance | $\pm 0.2$ µm | | | | | | |

To determine the $NRDP$ settings for robust design, we use the three design levels for each $NRDP$ as shown in Table 5.40. Since the fabrication material is always silicon, the Young's Modulus $E$ equals $146.9GPa$. In addition, the $RDP$ (pressure difference $p$) is assumed to be a sinusoidal pressure at a frequency of 100Hz. The amplitude of $p$ is limited in the range of $5,000$ to $15,000Pa$, the nominal design setting value is set to $10,000Pa$.

**Table 5.40** Design Levels for $NRDP$s (units: $\mu m$).

| $NRDP$s | $L$ | $b'$ | $h'$ | $h$ | $l_1$ | $b$ | $l_2$ |
|---|---|---|---|---|---|---|---|
| (-) | 1280 | 800 | 12 | 40 | 4 | 320 | 80 |
| Level (0) | 1600 | 1000 | 15 | 50 | 5 | 400 | 100 |
| (+) | 1920 | 1200 | 18 | 60 | 6 | 480 | 120 |

An exhaustive search to find optimal $NRDP$ setting for robust design is very difficult. The complexity of exhaustive search is $O(3^n)$, where $n$ is the number of $NRDP$s. However, most practical systems are dominated by some of the design parameters, and most higher order interactions are negligible. Therefore, a $1/3^p$ fraction of the original orthogonal array is used for experimental designs with reduced $O(3^{(n-p)})$ complexity, where $p$ is related to the order of interactions. We use the inner orthogonal array $L_8$ with two levels (-1, 1) for $NRDP$ tolerance, and the outer orthogonal array $L_{18}$ (Addelman-Kempthorne construction [92]) with three levels (-1, 0, 1) for $NRDP$s setting to directly evaluate the contribution of individual parameters to overall design robustness [31].

Therefore, by using the one-dimensional iterative *Fibonacci search* method, the setting points of $P_a$, $P_a(min.)$ and $P_a(max.)$, with the minimum flow rate and the maximum flow rate, respectively, can be obtained for each $NRDP$ setting. In addition, by calculating the average $SNR$ value at the $P_a$ nominal setting $P_a(nom.)$, $P_a(min.)$ and $P_a(min.)$, the average robustness of a setting of nonreconfigurable "hardware" platform is obtained. The design procedure is illustrated in Figure 5.30, and explained as follows.

1. *Design the outer array*
   Based on the orthogonal array $L_{18}$, and the three design levels for each $NRDP$ shown in Table 5.40, the outer array is obtained as in the Figure 5.30, each row represents a setting of $NRDP$s. For instance, in the first row, the $L$ value, $-1$, means that the length of the cantilever is $1280\mu m$ in this setting.

2. *Design the inner array*
   Depending on the $NRDP$ tolerance shown in Table 5.39 and the $L_8$ orthogonal array structure [92], we can obtain the inner array for each row of the outer array, meaning each setting of $NRDP$s. For example, if the inner array shown in Figure 5.30 is developed depending on the $i^{th}$ setting of the outer array, the value of $L$ at the first row in the inner array, $-1$, implies that the value of $L$ is $1599.8\mu m$.

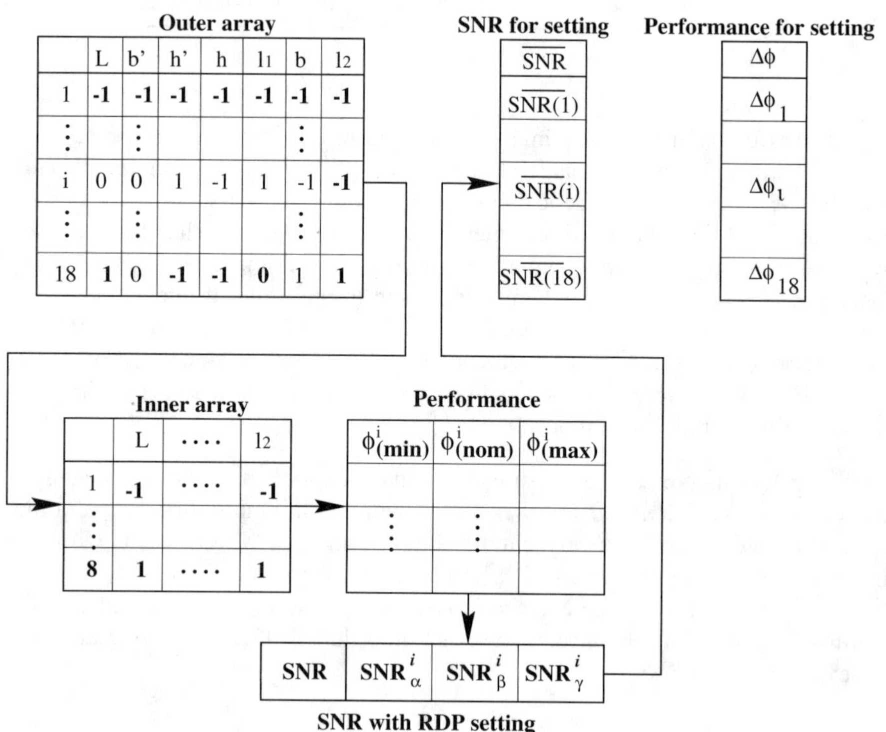


**FIGURE 5.30**
**Experiment design**

3. *Search the design performance*
Within the degree of "programming" of the $RDP$ (pressure difference $P_a$), we can obtain three performance values for $i^{th}$ $NRDP$ setting by the iterative searching method: minimum flow rate ($\Phi^i_{min}$), normal flow rate ($\Phi^i_{nom}$), and the maximum flow rate ($\Phi^i_{max}$).

**Table 5.41** Average $SNR$ Ratio for the Design Parameters.

| Parameter | Average $SNR^*$ by level | | |
|---|---|---|---|
| | lower (-1) | normal (0) | high (+1) |
| $L$ | 66.67 | 65.90 | 64.95 |
| $b'$ | 65.42 | 65.67 | 66.43 |
| $h'$ | 64.26 | 66.36 | 66.90 |
| $h$ | 65.86 | 65.87 | 65.79 |
| $l_1$ | 65.85 | 65.85 | 65.82 |
| $b$ | 64.47 | 65.97 | 67.08 |
| $l_2$ | 65.86 | 65.84 | 65.83 |

**Table 5.42** Flow-rate Range $\Delta\Phi$ $[\mu l/min]$for the Design Parameters.

| Parameter | $\Delta\Phi = \Phi_{max} - \Phi_{min}$ by level | | |
|---|---|---|---|
| | lower (-1) | normal (0) | high (+1) |
| $L$ | 218.26 | 272.64 | 326.58 |
| $b'$ | 218.54 | 272.53 | 326.42 |
| $h'$ | 272.50 | 272.80 | 272.18 |
| $h$ | 272.42 | 272.49 | 272.57 |
| $l_1$ | 272.47 | 272.50 | 272.51 |
| $b$ | 217.83 | 272.47 | 327.18 |
| $l_2$ | 272.33 | 272.49 | 272.66 |

4. *Calculate the robustness for each setting of design parameters*
Based on the $SNR$ objective functions, the related system robustness for three design performances can be calculated as $SNR^i_\alpha$, $SNR^i_\beta$, and $SNR^i_\gamma$, respectively. The overall robustness of a setting of design performance ($\overline{SNR^i}$) is the average of each $SNR$s, as given in (5.46).

$$\overline{SNR^i_\alpha} = \frac{\sum_{i=1}^{8} SNR^i_\alpha}{8}$$

$$\overline{SNR^i_\beta} = \frac{\sum_{i=1}^{8} SNR^i_\beta}{8}$$

$$\overline{SNR^i_\gamma} = \frac{\sum_{i=1}^{8} SNR^i_\gamma}{8}$$

5. *Calculate the main effect*
   As shown in the Table 5.41, the main effect for design levels of each design parameter is the average of the $SNR$ with the same setting for the whole design solutions. For example, the main effect for the length of the cantilever $L$ at lower level $(-1)$, $M_L^{-1} = 66.67$, is the average of $SNR$ for the design solutions where $L$ is set to $-1$.

6. *Calculate the flow-rate for design levels*
   The application flexibility of the system, the range of the flowrate for each design parameter setting can also be calculated, as shown in Table 5.42. The $\Delta\Phi$ is the difference between the maximum flowrate and the minimum flowrate within the *Performance* table in Figure 5.30. For example, regardless of other design parameter settings, with $L$ set at the lower level $(1280\mu m)$, the system flow-rate range $\Delta\Phi$ is $218.26\mu l/min$

Additionally, the following important observations can be made concerning the optimal system design:

- Figures 5.31 and 5.32 illustrate that the microvalve robustness and the flow-rate range for different $NRDP$ setting within the range of $RDP$s, respectively. The setting of $NRDP$s is dependent on the design objectives (robustness versus performance range); there does not exist a unique design point that satisfies conflicting design requirements.

- In studying robust design, we note that the length of the cantilever $(L)$, the width of the cantilever $(b')$, the thickness of the cantilever $(h')$, and the width of the valve seat $(b)$ have a significant effect on $SNR$. Except $h'$, they also have a significant effect on the average flow-rate range. The setting with $L_{(-1)}$, $b'_{(+1)}$, $h'_{(+1)}$, $h_{(0)}$, $l_{1\,(0)}$, $b_{(+1)}$, $l_{2\,(-1)}$ is clearly the most robust. The robustness of this setting of the design parameters, $\overline{SNR}$, is the average of the main effects for each design parameter.

- In attempting design for wider flow-rate range design, we note that the length of the cantilever $(L)$, the width of the cantilever $(b')$, and the width of the valve seat $(b)$ have a significant effect on the average flow-rate range. The setting with $L_{(+1)}$, $b'_{(+1)}$, $h'_{(-1)}$, $h_{(+1)}$, $l_{1\,(-1)}$, $b_{(+1)}$, $l_{2\,(+1)}$ possesses the widest flow-rate range.

Figure 5.33 and Table 5.43 present the optimal design results: the optimal design for the widest flow-rate range, and the optimal design for the robustness. In addition,

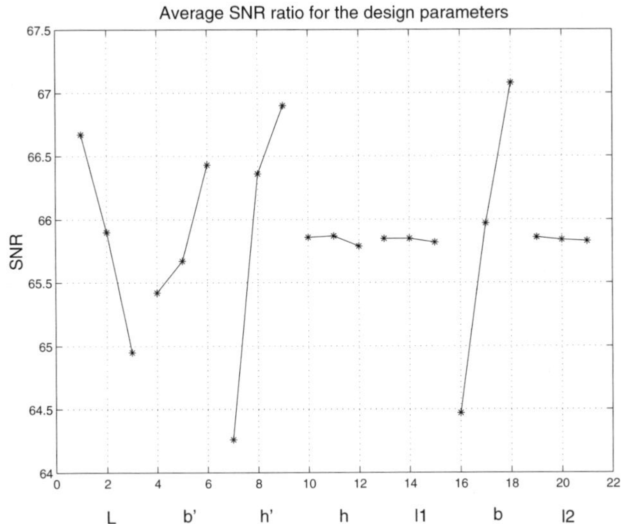


**FIGURE 5.31**
**Plot of design parameter effect on SNR. $L$, $b'$, $h'$, $b$ show the significant effect on SNR.**

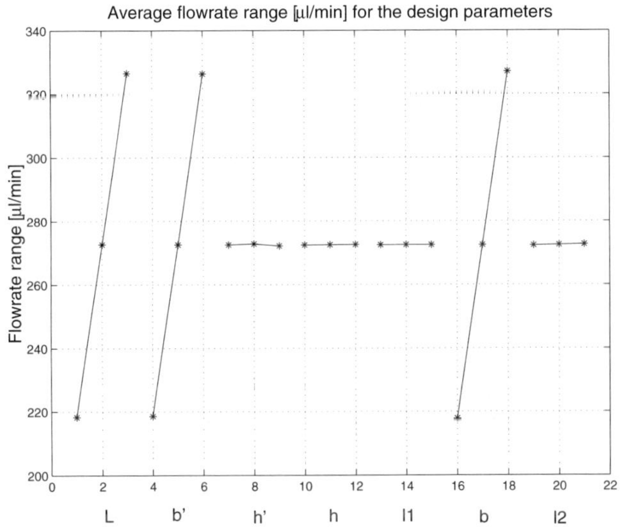


**FIGURE 5.32**
**Plot of design parameter effect on the flow-rate range. $L$, $b'$, $b$ show the significant effect on the flow-rate range.**

Figures 5.31 and 5.32 also directly provide very useful information for related performance improvement. For example, increasing the value of $L$, $b'$ and $b$ increases the range of flow rate, while decreasing the value of $L$ and increasing the value of $b$ improves the microvalve robustness. Based on these performance analyses, other feasible design solutions can also be obtained.

**Table 5.43** Design Results

| | $L$ | $b'$ | $h'$ | $h$ | $l_1$ | $b$ | $l_2$ | Range ($\Delta\Phi$) | $(\overline{SNR})$ |
|---|---|---|---|---|---|---|---|---|---|
| Nominal Design | 1600 | 1000 | 15 | 50 | 5 | 400 | 100 | 272.86 | 66.66 |
| Most Robust | 1280 | 1200 | 18 | 50 | 5 | 480 | 80 | 314.00 | 69.13 |
| Widest Range | 1920 | 1200 | 15 | 60 | 6 | 480 | 120 | 469.86 | 63.45 |

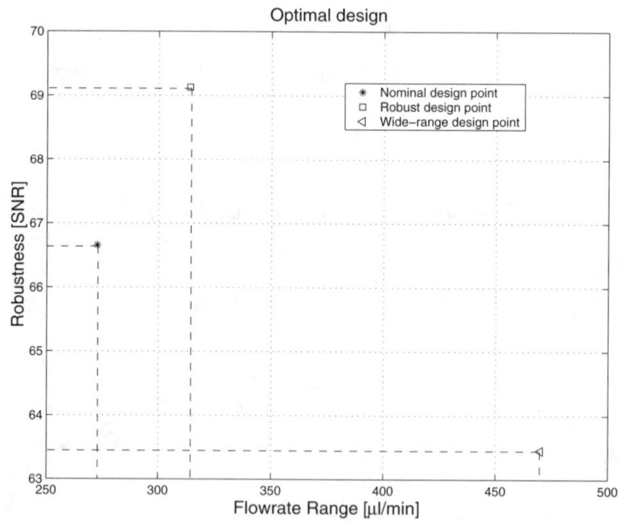


**FIGURE 5.33**
**Plot of optimal design points: the nominal design point, the optimal design for the widest flow-rate range, and the optimal design for the robustness.**

We have leveraged the principles of the hardware/software co-design methodology to develop a novel design approach for enhancing the application flexibility of composite microsystems. This methodology is also useful for rapid prototyping and CAD tool development. The future work is to extend the proposed co-design approach to higher-level design optimization, including system architecture and performance evaluation.

## 5.6 Conclusion

In this chapter, several simulation-based design and process optimization algorithms were developed. The more efficient simulation methodologies, the bootstrap method and the factorial design method, are introduced. Case studies on a comb drive microresonator and a microvalve illustrate that these simulation design methods can reduce the computational cost while providing good parameter estimates. Then, a validation strategy is discussed. The objective verification method is used to verify the optimization results. To make the design solution match the design performance requirement efficiently, an optimal on-target design algorithm was developed based on a statistical response analysis strategy. A design for a microvalve is studied to illustrate this methodology. In addition, by investigating Taguchi experimental design and statistical process control methods, we demonstrated a robust design methodology. Two examples are given for the robust design, an electrostatic-comb microresonator and a microvalve. Moreover, on the analogy to a hardware/software co-design methodology, a novel application-flexibility design methodology is demonstrated. This method can leverage the design for additional applications based on the Taguchi robust design and the response surface methodologies. A case study of a microvalve is presented.

# Chapter 6

## Performance Evaluation

Microfluidic devices that combine existing components into a system are now being routinely developed [17], [16]. These systems are usually based on continuous-flow components, such as microvalves, micropumps and channels. There are several problems inherent in such designs. These problems include complex system architecture, dead-volumes, and integration complexity. On the other hand, new microfluidic systems may be based on droplet actuation, which uses electrowetting-based actuation to move fluidic samples. Droplet-based technology has been proposed for microreactor application [104], and chemical analysis [105].

In this chapter, a performance comparison is presented between two types of microfluidic systems—continuous-flow systems and droplet-based systems. The comparison is based on a specific microfluidic application—a polymerase chain reaction (PCR) system. The modeling and simulation of PCR are based on the SystemC design environment previously discussed. Section 6.1 introduces the principles of PCR amplification, as well as the detection of amplification and the purification of the PCR product. Section 6.2 discusses three continuous-flow PCR systems and their physical implementation. These continuous-flow PCR systems include a sequential, continuous-flow PCR system [7], a detectable PCR system [8], and a reusable continuous-flow PCR system. This reusable PCR system is based on a reconfigurable microliquid handling system architectural design [76]. Section 6.3 presents a PCR system based on droplet technology. Its physical implementation is also proposed. The comparison between these four types of microfluidic systems is discussed in Section 6.4. The comparison is based on the system design complexity and system performance. In Section 6.5, we present an architectural design and optimization methodology for performing biochemical reactions using two-dimensional electrowetting arrays.

## 6.1 Introduction

In this section, the principle of PCR is introduced. Several important functional units are also studied using typical values from the literature.

### 6.1.1 Polymerase Chain Reaction (PCR)

Since the first report of specific DNA amplification using the polymerase chain reaction (PCR) in 1985 [106], PCR has become a powerful tool for the detection of pathogens, where the amount of sample is often small and direct detection would be impossible. Its specific application is for rapid enzymatic amplification of specific DNA fragments. PCR can amplify genomic DNA exponentially using temperature cycles. Considering a DNA duplex consisting of regions *ABCDE*, as shown in Figure 6.1, if the sequences of $B$ and $D$ are known, then millions of copies of $C$ (the target) can be obtained by the PCR. One strand of this duplex is denoted with $a-b-c-d-e$, and the complementary strand is denoted with $a'-b'-c'-d'-e'$. PCR is carried out by adding the following components to a solution containing the target sequence: (1) a pair of primers, $b$ and $d'$; (2) all four deoxyribonucleoside triphosphates (dNTPs); and (3) a heat-stable DNA polymerase. As shown in Figure 6.1 [106], a PCR thermal cycle consists of three steps:

1. *Strand separation.*
   The two strands of the parent DNA molecule are separated by heating the solution to $95^o$C for 45 seconds.

2. *Hybridization of primers.*
   The solution is then abruptly cooled to $54^o$C to allow each primer to hybridize to a DNA strand. Primer $b$ hybridizes to $b'$ on one strand, and primer $d'$ hybridizes to $d$ on the complementary strand. This annealing process takes 30 seconds.

3. *DNA synthesis.*
   The solution is then heated to $72^o$C, the optimal temperature for $Taq$ DNA polymerase. This allows the $Taq$ DNA polymerase to attach at each priming site (where primers have annealed) and extend (synthesize) a new DNA strand. Elongation of both primers occurs in the direction of the target sequence because the $3'$ end of primer $d'$ faces $c$, and the $3'$ end of primer $b$ faces $c'$. One of the new DNA strands is $b-c-d-e$ and the other is $a'-b'-c'-d'$. The process takes 90 seconds.

PCR can successfully amplify target DNAs for mutation analysis, genetic mapping and sequencing [106]. However, before subsequent analysis of the amplicon, the target DNA concentration of the PCR product has to be detected, and the PCR product has to be purified to remove unwanted salts, primers and enzymes.

### 6.1.2 PCR Detection for DNA Concentration

An attractive method for PCR is based on conductivity detection. The analytical response ($G$, conductivity) is related to the concentration of species [5], i.e.

$$G = \frac{(\lambda_+ + \lambda_-)C}{1000K} \tag{6.1}$$

where $\lambda_+$ and $\lambda_-$ are the limiting ionic conductances of cations and anions in solution, respectively. $C$ is the concentration and $K$ is the cell constant calculated from $L/A$, where $L$ is the distance between the electrode pairs, and $A$ is the electrode

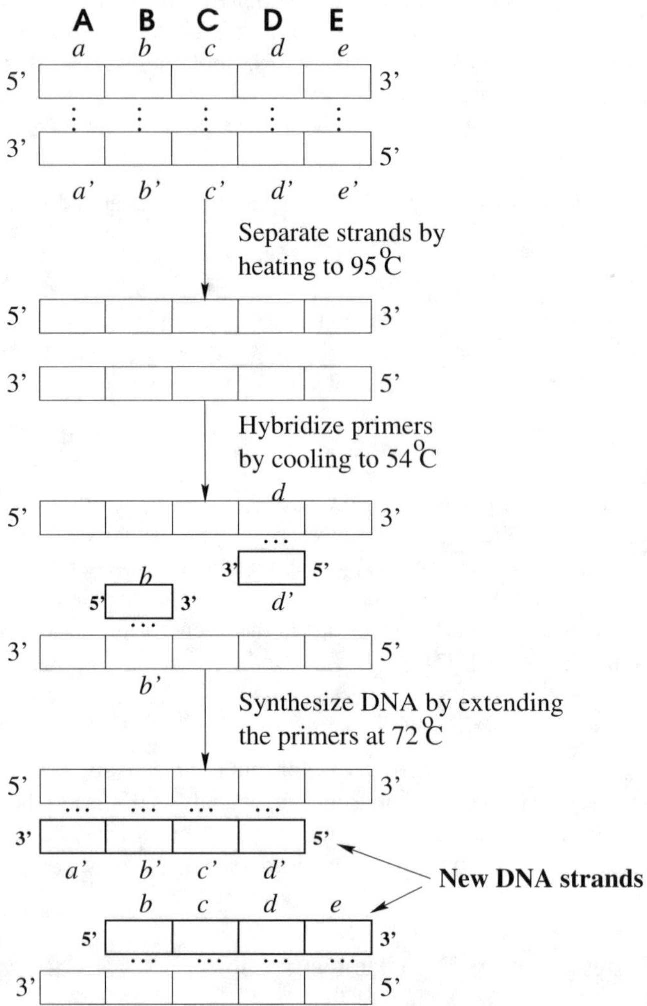


**FIGURE 6.1**
**Polymerase chain reaction (PCR). A cycle consists of three steps: strand separation, hybridization of primers, and extension of primers by DNA synthesis.**

area. As long as the analyte has a conductance different from that of the carrier solution, the DNA can be analyzed in its native state [5]. Conductivity can potentially offer several advantages compared to other common detection schemes used for DNA detection, such as UV absorbance or fluorescence [8]. These latter methods suffers from poor detection limits. The advantage of conductivity detection is that it is amenable to small column detection. Figure 6.2 shows the schematic of the conductivity cell [5]. The conductivity cell is constructed from Pt wires [107], where the diameter of each wire is $360\mu$m. The wires are inserted into opposite ends of a glass tee, the spacing distance between them is $60\mu$m. Two capillaries, $75\mu$m in diameter, are inserted into the tee. The bipolar-pulsed technique for conductance measurements is used for obtaining the solution conductivity [108]. The frequency of the pulses is typically 5000Hz. A flow-injection system needs to be built to conduct the solution to the conductivity detector. The flow rate ranges from one to several microliters per minute.

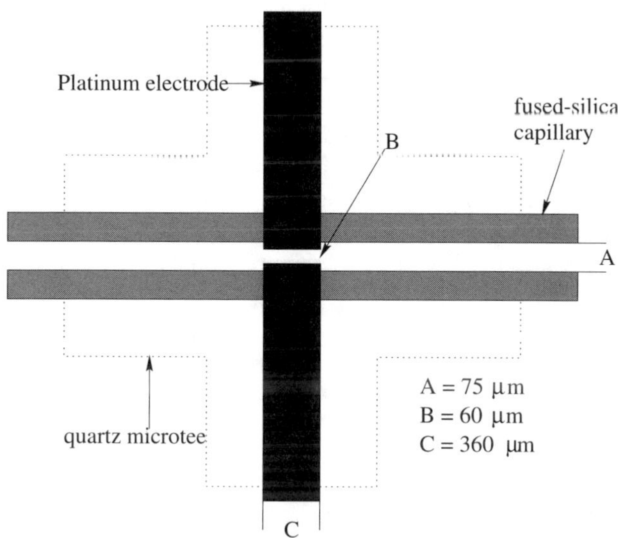


**FIGURE 6.2**
**Schematic of the conductivity cell [5]. The conductivity cell is constructed from Pt wires, quartz microtee and fused-silica capillary.**

### 6.1.3 PCR Purification

After amplification and conductivity detection, the DNA product needs to be purified to remove unwanted salts, primers, and enzymes for future analysis. There are several strategies proposed to perform purification of PCR-generated DNA. The purification technique used here is a reversed-phase separation, which can readily purify PCR products with high recoveries [5]. For most reversed-phase separations of DNA amplified via PCR, the ion-pairing agent is used. The common ion-pairing agents are tertiary or quaternary ammonium salts, for instance triethylammonium acetate or tetrabutylammonium phosphate. The chromatography of the PCR product is carried out using ion-pairing reversed-phase liquid chromatography (RPLC) with the mobile phase consisting of acetonitile-water (pH 7.0) and 50mM triethylammonium acetate. The full description of the PCR reverse-phase purification can be found in [109]. The volumetric flow ranges from one to several microliters per minute.

### 6.1.4 Acquisition Assumption

The PCR system can process a series of DNA solutions. The DNA solutions are sequentially moved into the system, amplified, detected, and purified, At the end, the processed DNA solutions are moved out for future processing. Without loss of generality, the volume for each DNA solution is assumed to be the same and equal to $30\mu l$. In addition, the acquisition function, $f(x)$, for the DNA solution to the PCR system is modeled by a traffic of liquid samples separated by interarrival times, denoted by $\{t_1, t_2, ...\}$. These interarrival variables are independent, identically distributed (IID) random variables, and they are characterized by an exponential probabilistic distribution given by (6.2), with a mean value of 4 minutes. That is $\lambda = \frac{1}{240}$ (Note the basic system time unit is a second). The incoming DNA solution is moved into the system until the associated system resources are available.

$$f(x) = \lambda e^{-\lambda x} \qquad x \geq 0 \tag{6.2}$$

## 6.2 Continuous-flow PCR System

Continuous-flow MEFS [11] use pressurized flow with mechanical flow control devices such as microvalves and micropumps. Continuous-flow MEFS can be integrated using two methods: (i) hybrid methods, where the elements are fabricated on different chips and set on a common substrate having microchannel interconnection; (ii) monolithic methods, where every element is fabricated on a single chip. In practice, most systems consist of a combination of these two forms. In this section, a three-way microvalve is introduced in Section 6.2.1. In addition, three continuous-flow PCR systems are discussed from Sections 6.2.2 to Section 6.2.4, respectively. Their physical implementations are also presented.

### 6.2.1 Three-way Microvalve

The three-way microvalve is one of most important parts for continuous-flow PCR system implementation. The structure of an advanced three-way microvalve is shown in Figure 6.3 [6]. The area of this three-way microvalve is about 8.5mm × 4.2mm. The valve consists of two mechanically fixed parts: a channel part and an actuator part. A pneumatic actuator is chosen because of its large displacement and high available pressure. The dimensions of the internal microchannel are 800$\mu$m in width, 8mm in length, and 30$\mu$m in thickness [110]. The typical flow rate in the microvalve ranges from one to tens of microliters per minute. The flow channel is divided into three zones having one inlet/outlet port, each of which is driven by the pneumatic actuator. The channel part has an oval cavity and three through-holes. The actuator part has right and left C-shaped chambers, and a center oval chamber. There are three different operation modes, as shown in Figure 6.4. In the first mode, the right chamber is actuated so that the right hole is closed, leaving a flow path between the left hole and the center hole. In the second mode, the center chamber is actuated, thus the center port is closed and a flow path is opened between the left and right ports. In the third mode, the left chamber is actuated. Therefore, the center port and the right port are connected along the channel, while the left hole is closed. Since each zone of the flow channel is covered entirely with the flexible membrane, dead volume is minimized and a good seal is obtained.

Table 6.1 shows the design parameters for a state-of-the-art three-way microvalve.

**Table 6.1** Three-way Microvalve Design Parameters and Their Nominal Values

| Parameters | Values |
|---|---|
| Area size | 8.5mm × 4.2mm |
| Internal channel width | 800 $\mu$m |
| Internal channel thickness | 30 $\mu$m |
| Internal channel length | 8 mm |
| Flow rate | 1 ~ 60 $\mu$l/min |

### 6.2.2 Sequential Continuous-flow PCR System

The continuous-flow PCR system is based on a single channel passing repetitively through the three temperature zones, as shown in Figure 6.5 [7]. The pattern of the chip layout determines the relative time that a fluid element is exposed to each temperature zone. In addition, the chip layout also defines the number of temperature

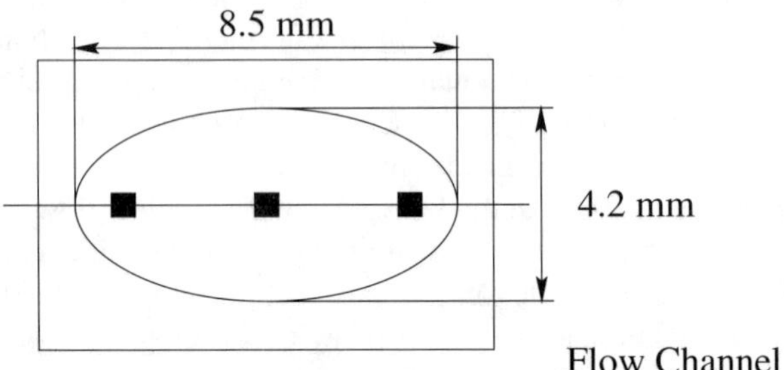


(a) Bottom View of the Channel Part

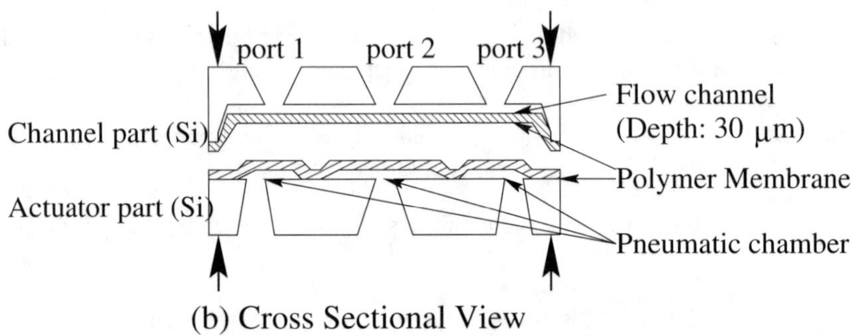


(b) Cross Sectional View

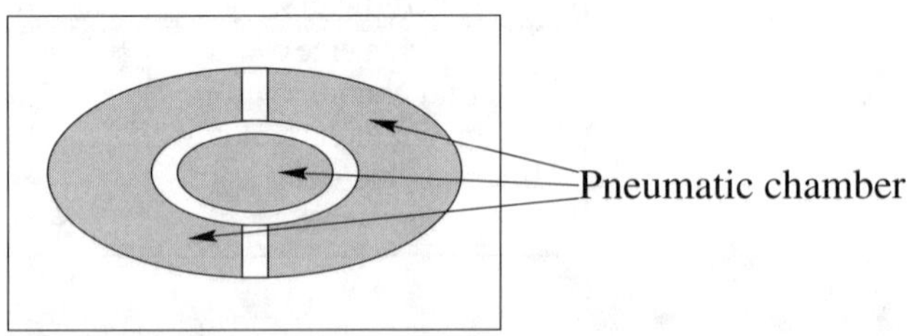


(C) Top View of the Actuator Part

**FIGURE 6.3**
**The three-way microvalve consists of a flow channel and a pneumatic actuator [6].**

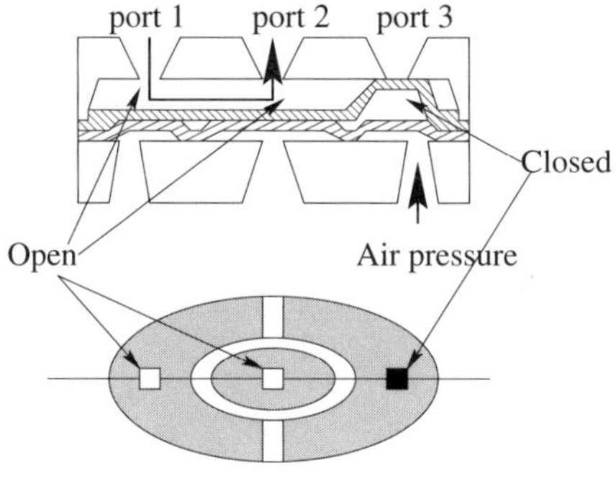


(1) Mode one

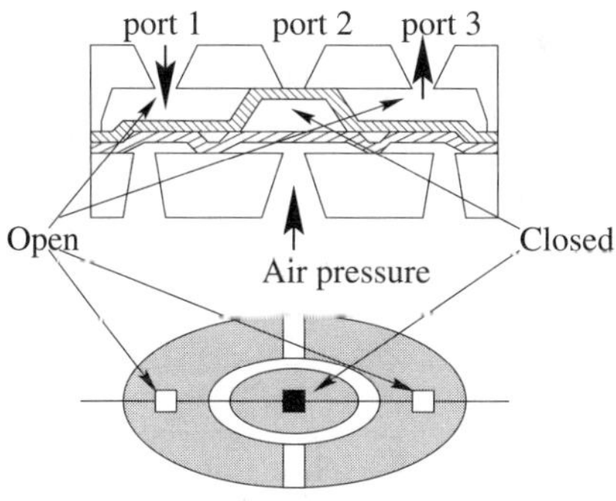


(2) Mode two

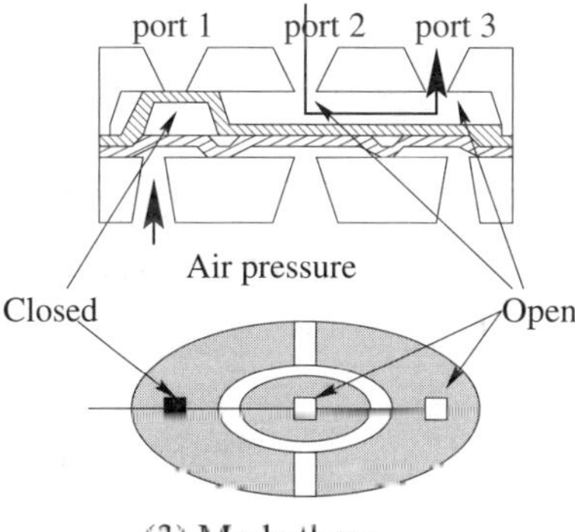


(3) Mode three

**FIGURE 6.4**

**Operation modes of a three-way microvalve consists of three modes: the left port connecting with the right port, the center port connecting with the right port, and the center port connecting with the left port [6].**

cycles, $n$, performed per run through the chip. The theoretical DNA amplification factor is $2^n$. We assume the use of a chip that generates 20 identical cycles, each having a time ratio of $4:4:9$ (melting : annealing : extension). Therefore, the amplification factor is $2^{20}$. Two of three inlets of the chip are used for a continuous buffer-fluid flow and for sample injection (the third inlet was not used). All fluids are pumped by hydrostatic pressure. The dimensions of the microchannel for a existing chip [7] are 40$\mu$m in width, 2.2m in length, and 90$\mu$m in height, The whole flow-through time is 4 min for the 20 cycles.

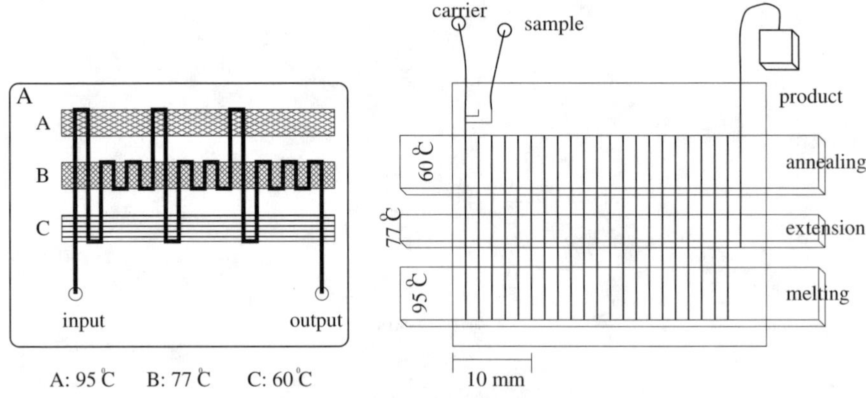


**FIGURE 6.5**
**Chip layout [7]. (A) Schematic of a chip for flow-through PCR. Three well-defined zones are kept at 95°C, 77°C, and 60°C. The channel passing through the three temperature zones defines the thermal cycling process. (B) Layout of the device. The device has three inlets on the left side and one outlet to the right.**

Figure 6.6 shows the sequential continuous-flow PCR system with a detection process and a purification process. The carrier pump brings a constant carrier flow through the system. The carrier solution can intermittently clean the processor, and it also is a carrier for the fluidic sample. A three-way microvalve [6] is used to switch the product flow to an analysis chamber or to a waste chamber.

Because these three process elements are connected to each other sequentially, the flow rate of each processor has to be the same. Based on the dimensions of the PCR chip, total volume of the microchannel, $V_{channel}$, is

$$\begin{aligned} V_{channel} &= W \times L \times h \\ &= 40\mu m \times 2.2m \times 90\mu m \\ &= 40\mu m \times 2200mm \times 90\mu m \\ &= 7.9\mu l \end{aligned} \tag{6.3}$$

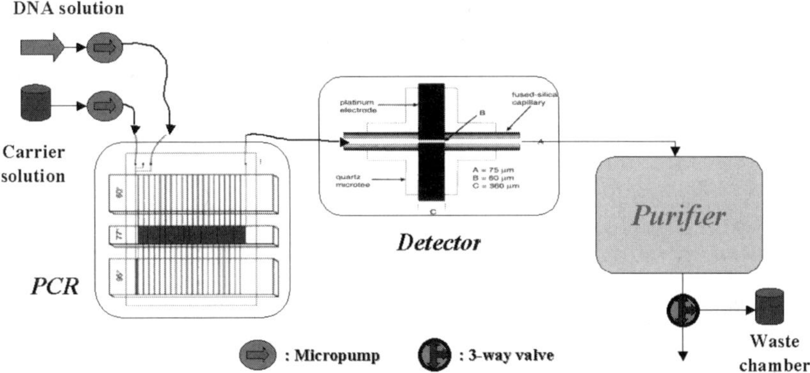


**FIGURE 6.6**
**Sequential continuous-flow PCR system consisting of three process elements: PCR, a detector, and a purifier. The carrier-flow pump brings a constant carrier flow through the system. A three-way microvalve is used to switch the product flow to an analysis chamber or to a waste chamber.**

where $W$ is the channel width, $L$ is the channel length, and $h$ is the channel height. Therefore, the flow rate $\Phi$ is as follows.

$$\Phi = V/t = 7.9\mu l/4min = 1.975\mu l/min \tag{6.4}$$

The distance between two processors, which is always several millimeters, is much shorter than the length of the microchannel of the PCR, 2.2 meters. Therefore, the transportation time between two processors can be neglected. Table 6.2 shows the critical processing time for a 30$\mu$l sample.

**Table 6.2** Design Parameters and Their Nominal Value for a Sequential Continuous-flow PCR System

| Parameters | Values |
|---|---|
| Processing time for PCR | $30/1.975 = 15$ min |
| Detection time | $30/1.975 = 15$ min |
| Purification time | $30/1.975 = 15$ min |

Figure 6.7 shows the system throughput. Figure 6.8 shows the system resource utilization for process elements: the PCR, the concentration detector, and the DNA purifier. Due to the sequential connection between these process elements, each pro-

cessor is under-utilized, and the system yield is low. In addition, the concentration of DNA solution can only be detected at the end of the PCR process. That is, the unsatisfied DNA solution cannot be fed back for reprocessing due to the uni-directional fluidic flow. The sequential, continuous-flow PCR system lacks the process-monitor capacity.

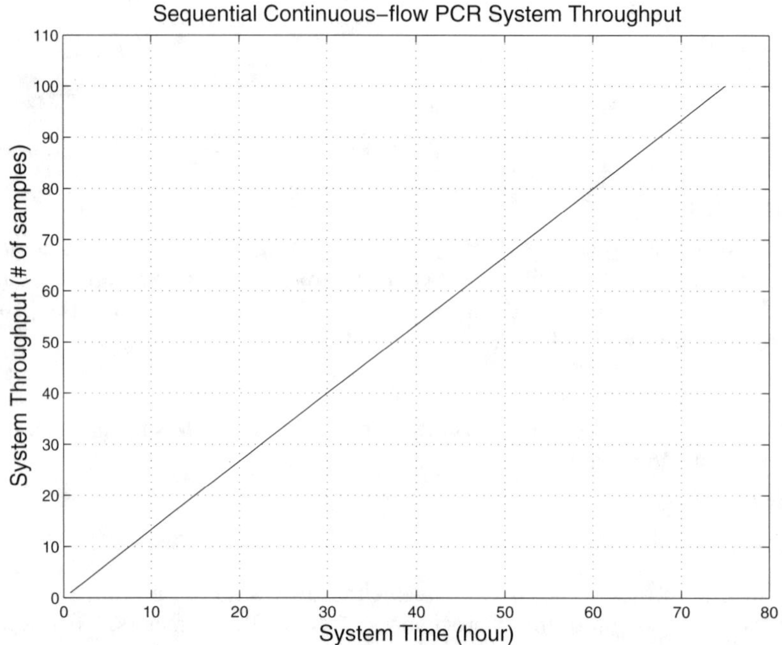


**FIGURE 6.7**
**System throughput of the sequential continuous-flow PCR system.**

### 6.2.3 Detectable PCR System

The detectable PCR system can continuously monitor the change of DNA concentration during the PCR process. The closed-chamber PCR chip with optical detection belongs to this category. Advantages of the PCR chip are its small size (6 mm $\times$ 4 mm $\times$ 1.5 mm) and the sealing of the chamber with a Pyrex wafer using the anodic bonding method. The transparent surface of the Pyrex makes it possible to incorporate optical readout methods [8].

Figure 6.9 shows the layout of a closed-chamber PCR chip [8]. The heaters and the temperature sensor are made of platinum 200 nm thick with an adhesion layer

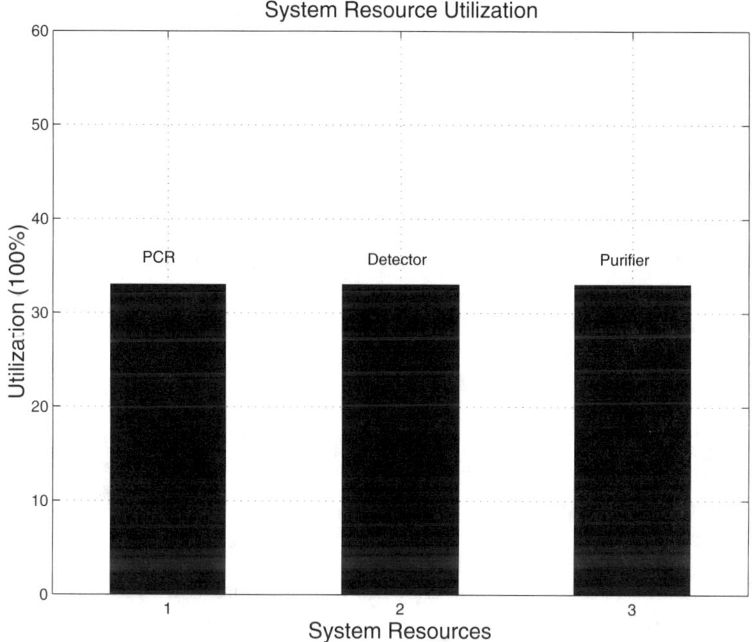


**FIGURE 6.8**
**System resource utilization of the sequential continuous-flow PCR system, each processor is under-utilized.**

of 20 nm of titanium. The sensor is designed to have four contact pads, and it is located under the bottom of the chamber. The two heaters are placed directly under the chamber side walls and can be connected in parallel or in series.

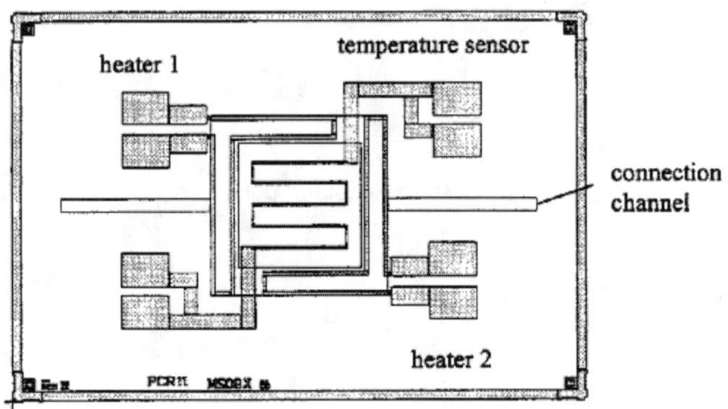


**FIGURE 6.9**
**Topview of the layout of a PCR chip. There are two heaters and one T-size temperature sensor [8]. © 2000 IEEE. With permission.**

Figure 6.10 shows the picture of a packaged PCR chip. The modified hypodermic needles of 0.6mm diameter are glued into the trenches of the Pyrex. The PCR solution can be conveniently introduced on the chip with a standard syringe or an injection pump.

In order to make optical detection more efficient, fluorescent dyes are added into the reaction mixture. This method is termed *homogeneous reporting chemistry* [8]. With these chemistries, fluorescence is detected using fluorimeters that are relatively simple. Fluorimeters collect the signal from the whole sample rather than at a specialized surface. The only requirement is that an optical window is available in the PCR chamber to allow the passage of visible light in and out of the reaction mixture. The pre-processor is needed to mix fluorescent dyes with DNA solution.

#### 6.2.3.1 Mixing Group

The basic structure of a mixing group is shown in Figure 6.11 [111]. The inlet unit connects the microsystem with the macroscopic environment. The micropump delivers a constant flow rate for liquids. There are two microvalves located at the input and the output of each micropump, which can prevent the reverse flow of the liquid. In addition, each micropump must connected with a flow sensor to measure

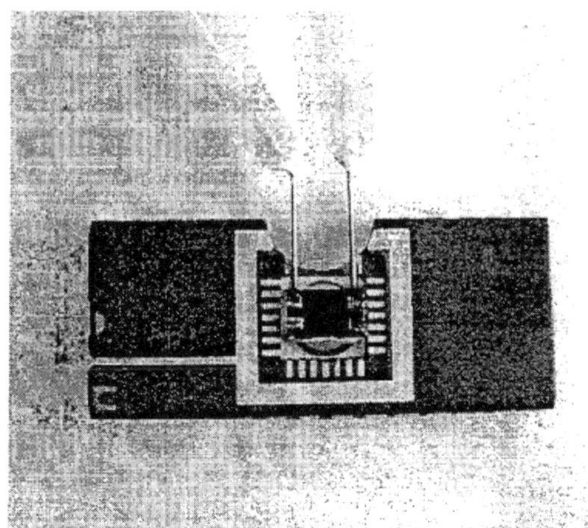

**FIGURE 6.10**
**Picture of a packaged PCR chip [8]. © 2000 IEEE. With permission.**

the flow rate [112]. [113] shows a flow sensor connected with a micropump based on the microfluidic circuitboard.

Due to the mixing process, the average flow rate for both liquids is about 10$\mu$l/hour. We assume the flow rate in our system is 10$\mu$l/min due to small volume of fluorescent dyes. The mixing of fluids in small channels is a technical challenge. In a very small channel, the liquid exhibits only laminar flow. A good mixing of different liquids requires turbulence in the flow. A simple solution is to coil the route of the microchannel in the mixer.

Table 6.3 shows the critical design parameters for the mixing group based on a state-of-the-art design.

**Table 6.3** Design Parameters and Their Nominal Value for a Mixer

| Parameters | Values |
|---|---|
| Flow rate | 10$\mu$l/min |
| Internal pump size | 7mm $\times$ 7mm [1] |
| Internal valve size | 2 mm $\times$ 2 mm [16] |
| Shut-off flow rate | $\leq$ 0.1$\mu$l/min |
| Mixing unit size | 1mm - 4mm |

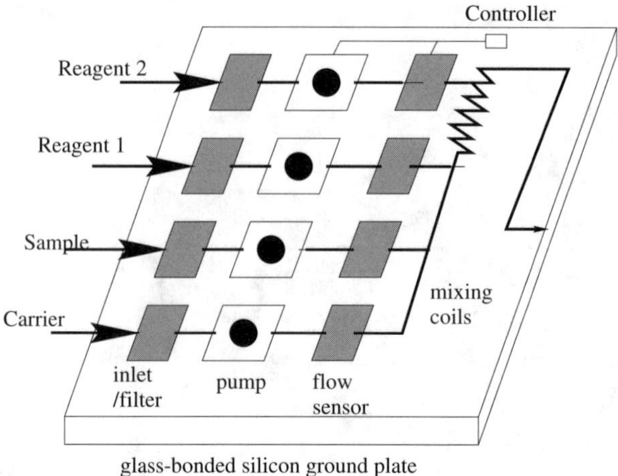


**FIGURE 6.11**
**The PCR mixer consists of inlet units, actuation pumps, flow sensors, and mixing coils.**

#### 6.2.3.2 Transportation Expense

Transportation time is a very important factor influencing the continuous-flow system performance. In contrast to the sequential continuous-flow PCR system, the closed-chamber PCR chip begins thermal cycling after the whole DNA solution is moved into the chamber. We have to be concerned about the transportation time from the mixer to the PCR chip.

Since the internal channel length of a three-way microvalve is 8mm, the viable length of the channel between two processors is approximately equal to the number of three-way microvalves multiplied by the length of the internal channel. For instance, there are two three-way microvalves between the mixing unit and the PCR chip, thus the channel length $L$ between them is at least 16mm. The regular cross-section area $A$ of the microchannel is $1600\mu m^2$ [114]. The volume of the channel between two processors is given by

$$Volume = A \times L = 1.6 \times 10^{-3} \times 16 = 0.0256\mu l \tag{6.5}$$

Although the maximum flow rate of a micropump can be 1000$\mu$l/min, the realistic flow rate $v_{liquid}$ is around 10$\mu$l/min due to the limitation of the three-way microvalve and the mixer flow rate. Therefore, depending on (6.5), the transportation time for the mixed liquid, whose volume $V$ is equal to 30$\mu$l, between the mixer and the PCR chip is

$$t = \frac{V}{v_{liquid}} = \frac{30\mu l + L \times A}{10\mu l/min} \simeq 3\ minutes \tag{6.6}$$

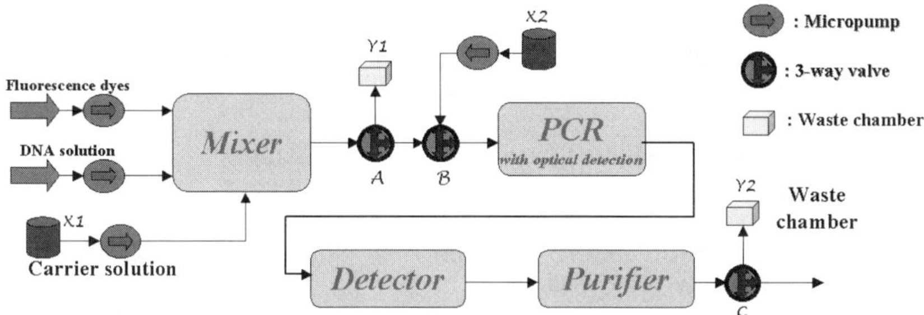


**FIGURE 6.12**
**Closed-chamber continuous-flow PCR system consists of four process elements: a mixer, a PCR, a detector, and a purifier. The mixing-liquid cycle consists of the mixer, the carrier chamber $x1$ and the associated pump, as well as the three-way valve $A$ and the waste chamber $y1$. The PCR-liquid cycle consists of the PCR/Detector/Purifier, the carrier chamber $x2$ and the associated actuation pump, three-way valves $B$ and $C$, as well as the waste chamber $y2$. Based on the mixing-liquid cycle, the DNA solution can be moved into the mixer. By controlling the three-way microvalves $A$ and $B$, the solution can be moved from the mixer into the PCR. Using the PCR-liquid cycle, the solution can be moved from the PCR through the detector, to the purifier and on to the outlet. The three-way microvalve $C$ can direct the unqualified solution into the waste chamber.**

where $L$ is the distance between the mixer and the PCR chip. $L \times A$ is too small to be counted when compared to the liquid volume $V$.

#### 6.2.3.3 Closed-chamber continuous-flow PCR System

Figure 6.12 shows the closed-chamber continuous-flow PCR system with a mixer, a PCR, a detector and a purifier. Carrier pumps bring a constant carrier flow through the system. Three-way microvalves are used to switch the flow direction. The three-way microvalves in the pre-processor switch the carrier solution or DNA solution from the previous process to the processor. Three-way microvalves in the post-processor switch the DNA solution to the next step analysis chamber or to the waste chamber.

Because the mixer performs an independent flow cycle, the mixing time $t_{mixing}$ of DNA solution with the reagent for a continuous-flow PCR system is determined by the mixer flow rate.

$$t_{mixing} = V_{solution}/\Phi_{mixer} = 30/10 = 3min \tag{6.7}$$

where the volume of the DNA solution $V_{solution}$ is $30\mu l$, and the mixer design flow rate $\Phi_{mixer}$ is $10\mu$l/min.

However, the three process elements (PCR, detector, and purifier) are connected sequentially, the flow rate of each processor has to be the same. Based on (6.4), the flow rate $\Phi$ of these processors is $1.975\mu$l/min.

Table 6.4 shows the critical processing time for a $30\mu$l sample.

**Table 6.4** Design Parameters and Their Nominal Value of Detectable Continuous-flow PCR System

| Parameters | Values |
|---|---|
| Mixing time | 3 min |
| Transportation time from the mixer to the PCR | 3 min |
| Processing time for PCR | 4 min |
| Detection time | $30/1.975 = 15$ min |
| Purification time | $30/1.975 = 15$ min |

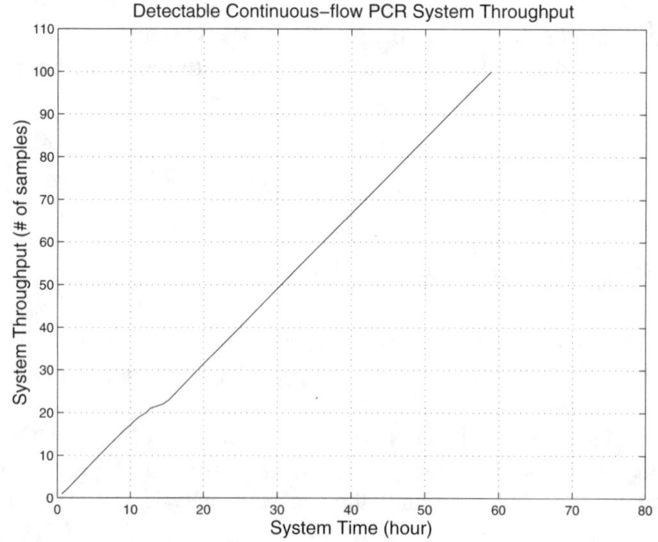


**FIGURE 6.13**
**System throughput of the detectable continuous-flow PCR system.**

Figure 6.13 shows the system throughput, and Figure 6.14 shows the system resource utilization for the mixer, the PCR, the concentration detector, and the DNA purifier. The detectable PCR system can detect the PCR product simultaneously with the PCR

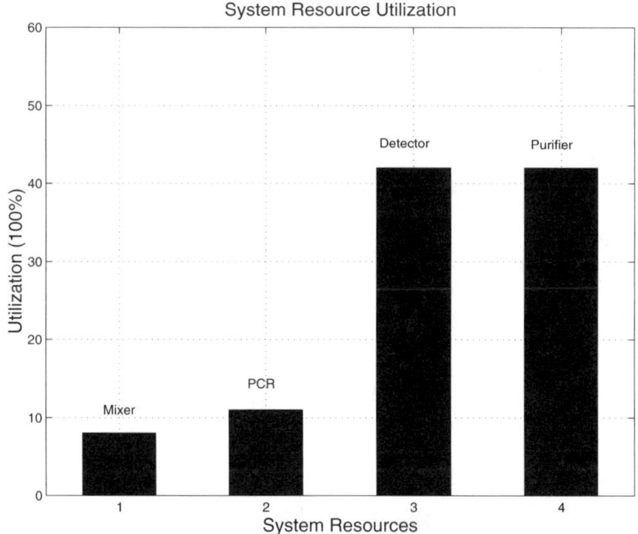


**FIGURE 6.14**
**System resource utilization of the detectable continuous-flow PCR system.**

processing. The DNA solution is moved out of the PCR only when the concentration of the target DNA matches the analysis requirement. The detectable PCR system reduces the PCR processing time, provides the process-monitor capacity, and improves system yield. However, due to the sequential connection between the process elements, each processor is still under-utilized.

### 6.2.4 Reconfigurable PCR System

In order to address the under-utilization of system resources, and improve the system yield, a reusable and reconfigurable PCR system is proposed, Its potential architecture is presented in Figure 6.15. The system is based on a reconfigurable microliquid handling system architectural design [76].

The design principle is based on the following factors.

- Independent Functional Groups
  As discussed previously, because of the sequential flow between microfluidic components—the PCR, the detector, and the purifier, the system performance is limited by the worst part of the system. For instance the conductivity detector's flow rate that can be $7.5\mu$l/min must be reduced to $1.975\mu$l/min to meet the PCR process requirement. Therefore, a redesigned architecture is required in which each functional group of components can be operated independently.

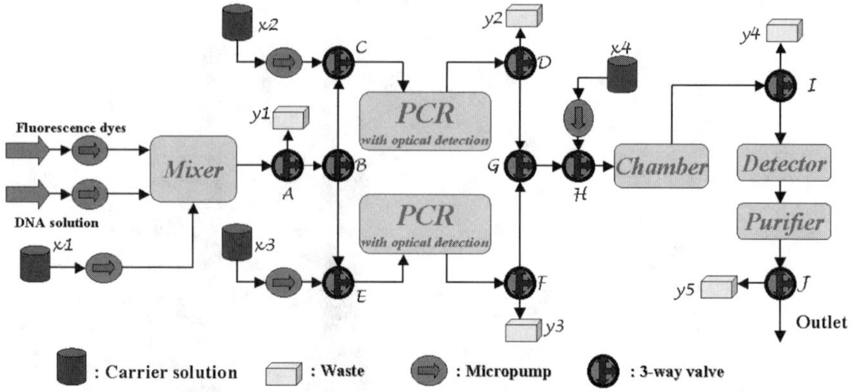

**FIGURE 6.15**
**Reconfigurable continuous-flow PCR system consists of three functional blocks: the pre-processing block, the processing block, and the post-process block. The combination of the carrier chamber $x1$ and the waste chamber $y1$ form the mixing liquid transportation path. Other fluidic paths include the PCR 1 cycle with the $x2$ carrier and the $y2$ waste chamber, the PCR 2 path with the $x3$ carrier and the $y3$ waste chamber, and the post-process path with the $x4$ carrier and the $y4$ or $y5$ waste chambers.**

- Independent Fluidic-Flow Cycle
  In order to make each functional block operate independently, an independent fluidic-flow cycle has to be built for each functional group. For instance, as shown in Figure 6.15, the combination of the carrier chamber $x1$ and the waste chamber $y1$ form the mixing fluid-flow cycle. This cycle not only carries the incoming DNA solution and associated reagent into the mixer, but it also carries the solution mixture from the mixer to a PCR chip.

- Temporary Storage Buffer
  Due to the performance difference between functional groups, temporary storage buffers are necessary between certain function groups. For example a chamber is inserted between the PCR chips and the detector shown in Figure 6.15. The conductivity process requires a flow rate of $7.5\mu$l/min. That rate is much less than the flow rate of the transportation channel between microfluidic components of $20\mu$l/min. Therefore, a chamber is necessary to temporarily store the PCR processed product. In addition, temporary storage buffers can separate the sequentially connected processors into several independently operating functional groups, thus benefiting system performance. For example this temporary chamber shown in Figure 6.15 can separate the processing chain into the processing block and the post-processing block. Therefore, the system performance has been improved from low throughput, as shown in Figure 6.13, to higher throughput shown in Figure 6.16.

- Reusability
  A fundamental facet of the reconfigurable architecture is that of reusability. Reusability implies a number of issues with regards to cross-contamination and cleanliness. The carrier flow is necessary for each functional group. When there is an incoming fluidic solution, the carrier flow takes the solution through the process. Between the interval of two fluidic solutions, the carrier flow with constant flow rate can function as a cleaning solution.

Table 6.5 shows the values of the design parameters for the reconfigurable continuous-flow PCR system. Based on (6.6), the transportation time from the mixer to the PCR chip is 3 minutes, while the transportation flow rate between the PCR and the chamber can be $20\mu$l/min. Therefore, the transportation time between the PCR and the chamber is 1.5 minutes.

**Table 6.5** Design Parameters for the Reconfigurable Continuous-flow PCR System

| Design Parameters | Values |
|---|---|
| Mixing time | 3 min |
| PCR time | 4 min |
| Detection time | 3 min |
| Purification time | 3 min |
| Transportation time between pre-processing block and the processing block | 3 min |
| Transportation time between the processing block and the post-processing block | 1.5 min |

Figure 6.16 illustrates the system throughput for a reconfigurable continuous-flow PCR system. Figure 6.17 shows the system resource utilization. There are 100 incoming DNA samples. The solution acquisition rate is modeled by an exponential probabilistic distribution, governed by (6.2). Nominal mean interarrival time is assumed to be 4 minutes, i.e. $\lambda = \frac{1}{240}$. Because of the limitation of UV absorbance and fluorescence [8], the backup detector using the conductivity technique may detect that the PCR product does not match the concentration requirement, this DNA solution is called an "unqualified solution". We assume that the amount of unqualified DNA solution is 10% of the total DNA solutions. The processed fluidic samples include qualified fluidic samples and unqualified fluidic samples.

The unqualified solutions are detected by the conductivity detector. The system correction capability is that if there is an unqualified solution detected by the conductivity detector, and one of the PCR units is available, the system can send this unqualified liquid back for additional thermal cycling. Because of the uni directional fluidic flow, the processed fluid sample cannot be sent back, and the reconfigurable

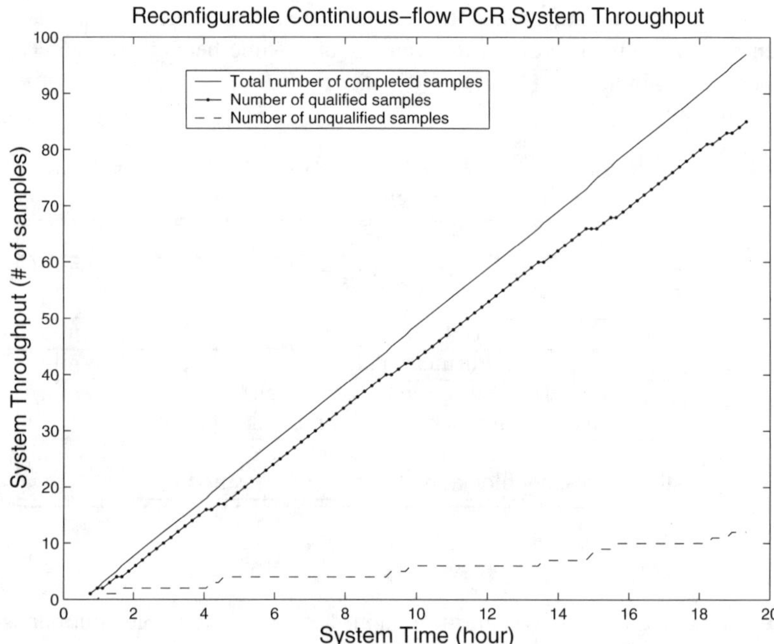


**FIGURE 6.16**

**System throughput of the continuous-flow PCR system. The total processed fluidic samples include qualified fluidic samples and unqualified fluidic samples.**

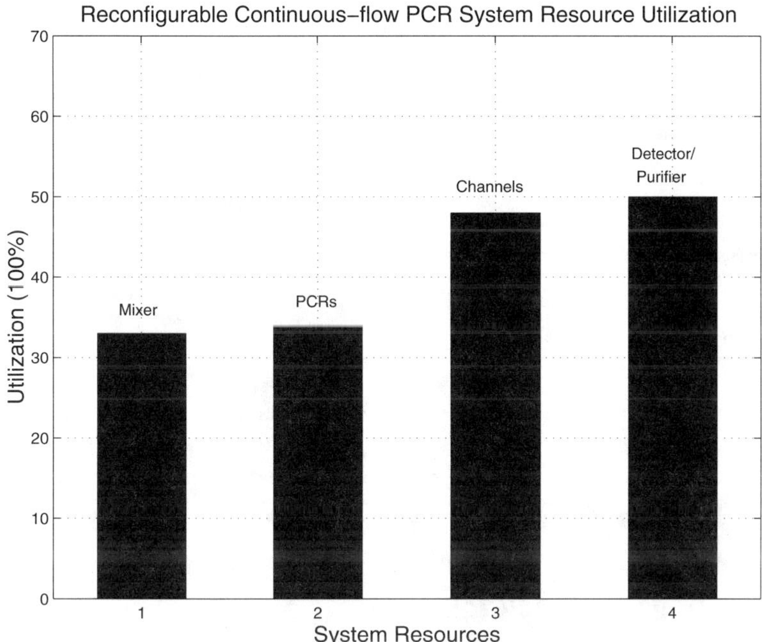


**FIGURE 6.17**
**System resource utilization of the continuous-flow PCR system. The flow control group is one of the system performance bottlenecks.**

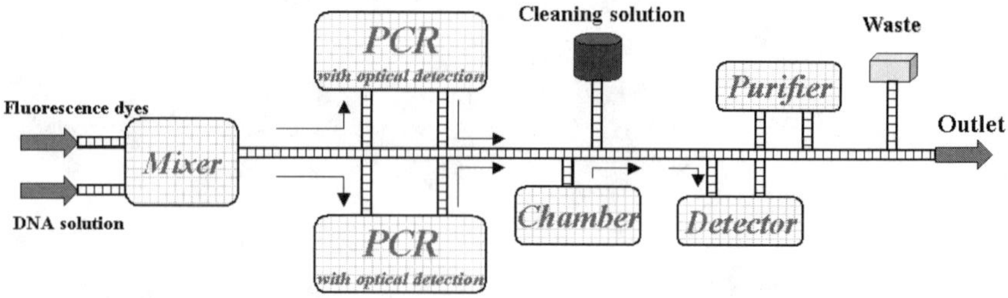

**FIGURE 6.18**
**Schematic view of the droplet-based PCR system.**

continuous-flow PCR system lacks a system-correction capability.

## 6.3 Droplet-based PCR System

Another physical implementation method for microeletrofluidic systems (MEFS) is based on droplet technology. It is based on electrically-driven liquid handling without mechanical elements. Section 6.3.1 introduces a droplet-based PCR, and its physical implementation is presented in Section 6.3.2.

### 6.3.1 A Droplet-based PCR System

The electrostatic actuation method has been proposed for manipulating microdroplet movement [104]. A potential architecture of a droplet-based PCR system is based on the electrowetting-based actuation presented in [105]. The rapid actuation of discrete liquid droplets is based on direct electrical control of their surface tension. The droplet-based PCR is presented in Figure 6.18.

The droplet-based PCR system has the same architecture as the reconfigurable continuous-flow PCR system. Both systems have the same fluidic processing blocks. The difference between them being the flow control subsystem. The DNA solution is moved into the PCR system through inlet units. The mixing units and other processor chambers are developed with electrode arrays. These arrays are used to control the fluidic flow. The transportation chain is used to connect different fluidic processing blocks. Because there are no big mechanical elements in the transportation chain, the distance between any two functional components is decreased. The direction of

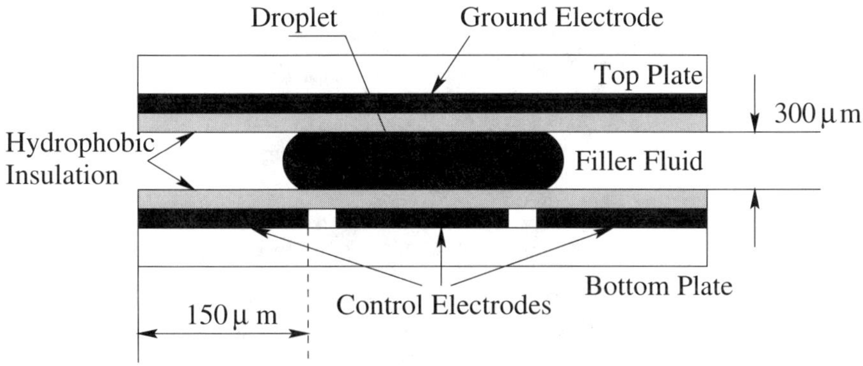


**FIGURE 6.19**
**Schematic cross-section of the electrowetting microactuator.**

movement of fluidic samples is reversible.

### 6.3.2 Physical Implementation

#### 6.3.2.1 Electrowetting microactuator

The electrowetting microactuator is presented schematically in Figure 6.19 [105]. A droplet of polarizable and conductive liquid is sandwiched between two sets of planar electrodes. The upper plate consists of a single continuous ground electrode, while the bottom plate consists of an array of independently addressable control electrodes. The distance between two sets of planar electrodes is $300\mu$m, and the edge of a control electrode is $150\mu$m. The control electrodes are square, as shown in Figure 6.20. The upper limit of the droplet speed is 15cm/second. With each electrode independently controlled, multiple fluidic droplets can be moved simultaneously. There must be at least one electrode between two droplets in order to maintain isolation.

Table 6.6 shows the critical design parameters for the electrowetting actuator.

**Table 6.6** Design Parameters for the Electrowetting Actuator

| Design Parameters | Values |
|---|---|
| Thickness between two planes | $300\mu$m |
| Width of control electrodes | $150\mu$m |
| Maximum droplet speed | 15 cm/second |

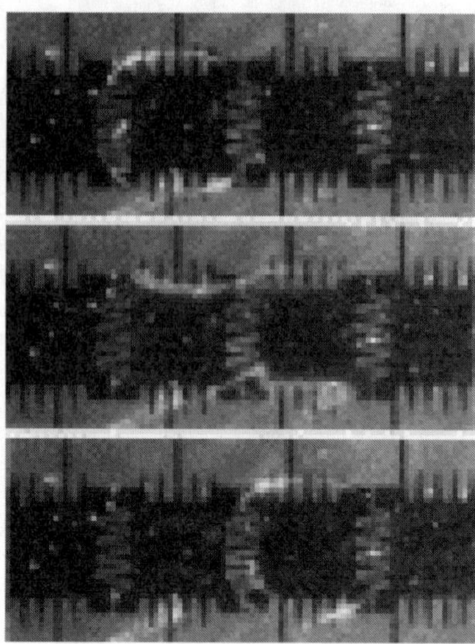

**FIGURE 6.20**
**Video frames of a moving droplet at 33 ms intervals [9].**

#### 6.3.2.2 Droplet Mixer

Figure 6.21 shows the concept of a "droplet mixer" [104]. Two sample droplets are fed from different inlet units, mixed with each other, and moved to the outlet. The droplet mixer concept offers the advantages of simple construction, no moving control devices, and no dead-volume. The concept of a mixer can be used for the droplet reactor design [104]. Because of the agitation during the droplet movement, the mixing time can be largely reduced. Normally the time for one $\mu$l droplet mixing is several seconds. Because electrodes can be controlled independently, large size droplet can be separated into several smaller droplets, and mixed simultaneously. Thus the mixing time of a droplet is not related to its volume. We assume that the mixing time for a 30$\mu$l sample is 10 seconds.

#### 6.3.2.3 Transportation Expense for Droplet-based PCR System

The topology of the grid array influences the route that a droplet must take in moving from one port to another. Hence the definition of the distance between two ports must consider the structure of the droplet-based PCR system. The distance between two ports is defined as the number of the electrodes between them, $n$. The number of droplets, $n_{droplet}$, whose volume $V_{droplet}$ is 1$\mu$l, existing in the 30$\mu$l liquid sample is

$$n_{droplet} = \frac{V_{liquid}}{V_{droplet}} = \frac{30\mu l}{1\mu l} = 30 \tag{6.8}$$

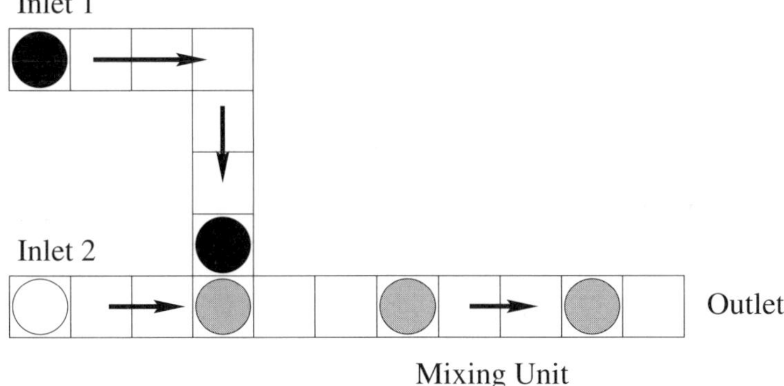


**FIGURE 6.21**
**Concept of the "droplet mixer." Two sample droplets are fed from different inlet units, mixed with each other, and moved to the outlet**

where, $V_{liquid}$ is the volume of a sample. Since the droplets can be moved simultaneously at a maximum speed of $v_{droplet}$ 15cm/s (which is equal to 8.2 electrodes/second, the width of the electrode is 1.82mm), the transportation time, $t$, of the liquid through $n$ electrodes is given by

$$t = (n + 3 + 2 \times (n_{droplet} - 2))/v_{droplet} \tag{6.9}$$

Since the distance between any two ports is within several millimeters, the number of electrodes, whose width is 1.82mm, between two ports can be assumed as 10. The transportation time of a 30$\mu$l sample between any two ports is approximately as follows:

$$\begin{aligned} t &= (n + 3 + 2 \times (n_{droplet} - 2))/v_{droplet} \\ &\simeq (10 + 3 + 2 \times (30 - 2))/8.2 \\ &\simeq \frac{69}{8.2} \\ &\simeq 8 \ seconds \end{aligned} \tag{6.10}$$

Table 6.7 shows the values of the design parameters for the droplet-based PCR system. We assume each droplet-based processor has the same processing time as that of the continuous-flow PCR system except the mixer.

#### 6.3.2.4 System Throughput

Figure 6.22 shows the system throughput of a droplet-based PCR system. Figure 6.23 shows the system resource utilization. There are 100 incoming DNA solutions, and their interarrival time is with an exponential probabilistic distribution,

**Table 6.7** Design Parameters for the Droplet-based PCR System

| Design Parameters | Values |
|---|---|
| Mixing time | 10 seconds |
| PCR time | 4 min |
| Detection time | 3 min |
| Purification time | 3 min |
| Transportation time between any two functional groups | 8 seconds |

the mean is 4 minutes. Because the direction of movement of fluidic samples is reversible, the droplet-based PCR system possesses the system correction capability. The unqualified solution detected by the conductivity detector may be sent back for additional thermal cycling when one of the PCR units is available.

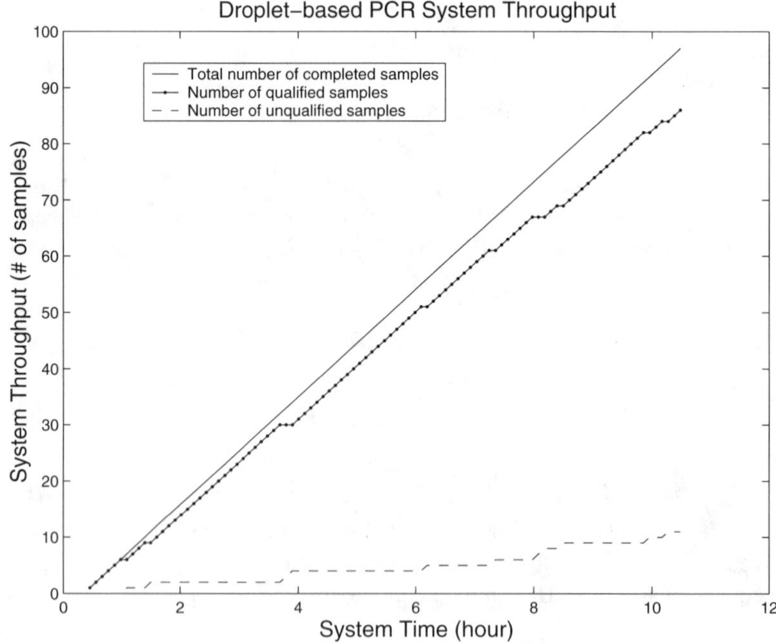


**FIGURE 6.22**
**System throughput of the droplet-based PCR system. The total processed fluidic samples include qualified fluidic samples and unqualified fluidic samples.**

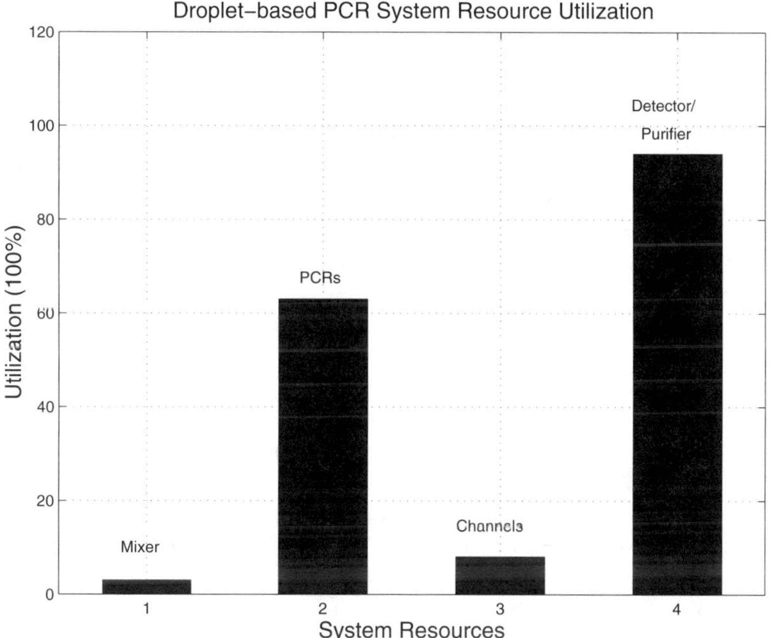


**FIGURE 6.23**

**System resource utilization of the droplet-based PCR system. The detector/purifier group and the PCR group are the performance bottlenecks. The utilization of the mixer and the transportation chain show their availability.**

## 6.4 Comparison between Continuous-flow PCR and Droplet PCR

In this section, we compare two types of PCR systems—continuous-flow PCR systems and droplet-based PCR systems. The evaluation is based on the system design complexity, system throughput, system resource utilization, and system correction capacity.

### 6.4.1 System Design Complexity

A reconfigurable continuous-flow PCR system needs a total of 16 mechanical flow control devices: 10 three-way microvalves and six actuation micropumps, as shown in Figure 6.15. In addition, each micropump must be connected to a flow sensor to measure the flow rate [112]. Five independent fluid-flow cycles dramatically increase the complexity of the system design and fabrication. Based on the basic design size for each device, the size of the flow control devices for the continuous-flow PCR is

$$Size \geq 10 \times (8.5 \times 4.2) + 6 \times (7 \times 7) \simeq 700mm^2$$

However, the droplet-based PCR system requires smaller and more convenient electronic control components. The fluidic flow is easy to control. Because the dimension of electrodes ranges from hundreds of micrometers to one millimeter, the size of flow control devices for the droplet-based PCR system is around several $mm^2$.

### 6.4.2 Performance Evaluation

#### 6.4.2.1 System Throughput

Figure 6.24 shows the system throughput comparison between the continuous-flow PCR system and the droplet-based PCR system. The droplet-based PCR system provides higher system throughput.

Table 6.8 shows the system throughput comparison between the four PCR systems described in this chapter. It demonstrates that the popularly used continuous-flow systems such as the sequential continuous-flow system and the detectable continuous-flow system have very poor system throughput. Although the reconfigurable continuous-flow system improves the system throughput, the slow transportation speed and the complex structure limits its wider application. The droplet-based system improves the system throughput, and also enhances the system yield.

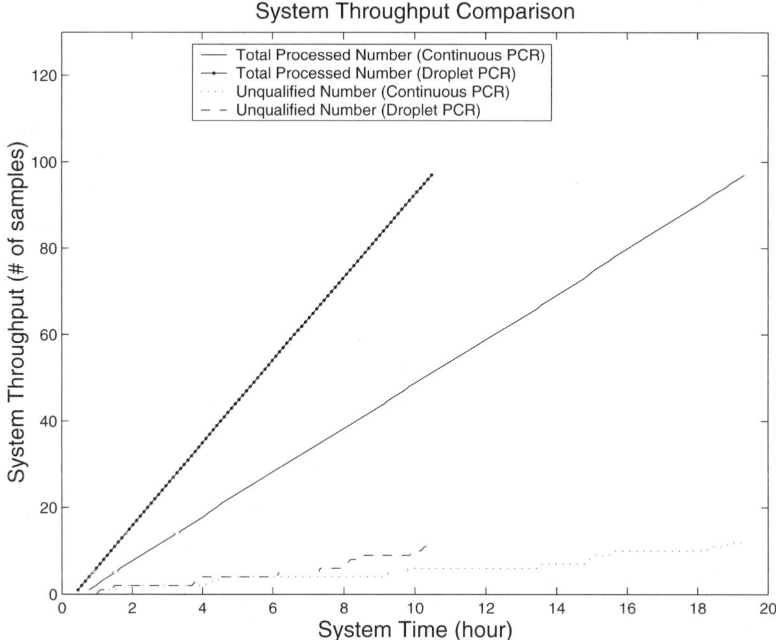


**FIGURE 6.24**
**System throughput comparison between the continuous-flow PCR system and the droplet-based PCR system. The droplet PCR system shows higher system throughput.**

**Table 6.8** System Throughput Comparison with 100 DNA Solutions

| PCR system | Finished time |
|---|---|
| Sequential continuous-flow PCR | 211820 |
| Detectable continuous-flow PCR | 270000 |
| Reconfigurable continuous-flow PCR | 69533 |
| Droplet-based PCR | 37730 |

#### 6.4.2.2 System Correction Capability

Because of the limitations of UV absorbance and fluorescence in analysis [8], the backup detector using the conductivity technique may determine that the PCR product does not match the concentration requirement. In contrast to the continuous-flow PCR system, there are no limitations for droplet flow movement. Therefore, if one of the PCR units is available, the unqualified liquid is sent back for another thermal cycling. System correction capability is an important measure of the system performance. Table 6.9 compares the system correction capacity between two systems when the interarrival time between fluidic samples is an exponential probabilistic distribution, the mean is 12 minutes. Droplet-based systems show very good correction capability.

**Table 6.9** System Correcting Capacity

| | Continuous-flow PCR | Droplet-based PCR |
|---|---|---|
| Total number of processed samples | 90 | 90 |
| Number of internal unqualified samples | 11 | 13 |
| Number of post-process samples | 0 | 10 |
| Number of qualified sample after post-process | 0 | 8 |
| Number of final unqualified samples | 11 | 5 |
| Unqualified sample percentage | 12.2% | 5.5 % |
| Correction Percentage | 0 % | 61.5% |

The continuous-flow PCR system and the droplet-based PCR system have nearly the same percentages of unqualified fluidic samples after reaction: 12.2% and 14.4%, respectively. The continuous-flow PCR system does not have the correction capability. However, in the droplet-based PCR system, 10 out of 13 unqualified DNA solutions go back for one more thermal cycling, and 8 out of 10 unqualified fluidic samples are corrected after re-processing. The correction percentage of the droplet-based PCR is 61.5%, and the final percentage of unqualified fluidic samples is reduced to 5.5%.

#### 6.4.2.3 System Processing Capacity

Acquisition rate (workload) is another important system-level design parameter influencing system performance. As discussed in Section 4.3, for a given architecture, the microfluidic system possesses a saturation processing capacity where resources are maximally utilized. Workloads less than saturation capacity under-utilize resources, whereas workloads greater than saturation capacity may decrease system quality. For instance, incoming fluidic samples have to wait longer if the system is saturated.

We assumed previously that the liquid sample acquisition rate is modeled by an exponential probabilistic distribution, governed by (6.2). Figure 6.25 shows the system processing capacity of the continuous-flow PCR system and the droplet-based PCR system, respectively. The acquisition rate ranges from $\frac{1}{700}$ to $\frac{1}{50}$.

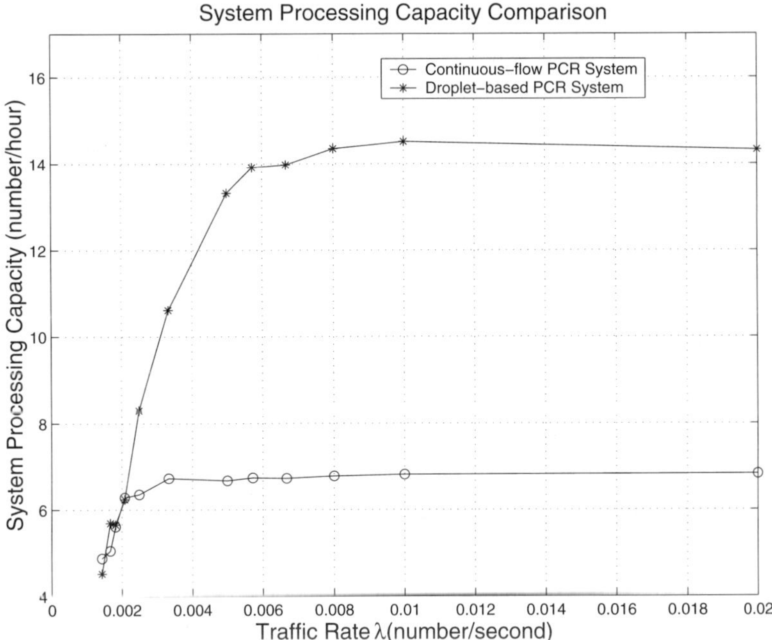


**FIGURE 6.25**
**System processing capability versus different traffic rate $\lambda$. After the system reaches the saturation, system processing capability (throughput) remains constant regardless of input rate variation.**

Figure 6.25 shows the system processing capability versus different traffic rates. The vertical axis presents the system processing capability, denoted by using the number of processed fluidic samples per hour. The horizontal axis represents the sample traffic rate, $\lambda$, meaning the number of fluidic sample arriving per second. When the sample traffic rate is low, the system throughput is nearly linear – the performance of the system approximates the ideal. At increased input rates, i.e. reduced interarrival time, the actual system processing capability increases and soon reaches saturation. Figure 6.25 shows that the droplet-based PCR system has higher processing capacity.

#### 6.4.2.4 Performance Enhancement

Figures 6.26 and 6.27 show the average time that each process takes in the continuous-flow PCR system and the droplet-based PCR system, respectively. The total time for each fluidic sample staying in the handling system consists of five periods: the mixing time, the PCR time, the delivery time, the analyzing time, and the time of waiting for system resources. Both figures show that the DNA solution spends too much time waiting for system resources. Figure 6.26 shows that the transportation part is performance bottleneck in a continuous-flow PCR system. Figure 6.27 shows that the PCR block and the detection/purification block are two principal problems for the droplet-based PCR system. As we mentioned previously, the droplet-based PCR system still uses the same processing blocks as the continuous-flow PCR system, the difference is only in the flow control blocks. Therefore, if the performance of the transportation block and the mixing block in the continuous-flow system do not improve, the performance of the droplet system is inherently better than that of the continuous-flow system. In addition, even though the improved architecture can enhance the continuous-flow system performance, it increases the system design and fabrication complexity.

## 6.5 Scheduling of Microfluidic Operations for Reconfigurable Two-Dimensional Electrowetting Arrays[1]

In the previous sections, we have presented a performance comparison between two types of microfluidic systems—continuous-flow systems and droplet-based systems. We have also demonstrated that the droplet-based microfluidic system provides higher performance, as well as lower design and integration complexity. In this section, we present an architectural design and optimization methodology for performing biochemical reactions using two-dimensional electrowetting arrays. We define a set of basic microfluidic operations and leverage electronic design automation principles for system partitioning, resource allocation, and operation scheduling. Fluidic operations are carried out through the electrostatic configuration of a set of grid points. While concurrency is desirable to minimize processing time, the size of the two-dimensional array limits the number of concurrent operations of any type. Furthermore, functional dependencies between the operations also limit concurrency. We use integer linear programming to minimize the processing time by automatically

[1]This section is based in part on "J. Ding and K. Chakrabarty and R. B. Fair, Scheduling of Microfluidic Operations for Reconfigurable Two-Dimensional Electrowetting Arrays. IEEE Transactions on Computer-Aided Design of Integrated Circuits and Systems, vol. 20, no. 6, pp. 1463-1468, Dec. 2001." © 2001 IEEE. Reprinted by permission.

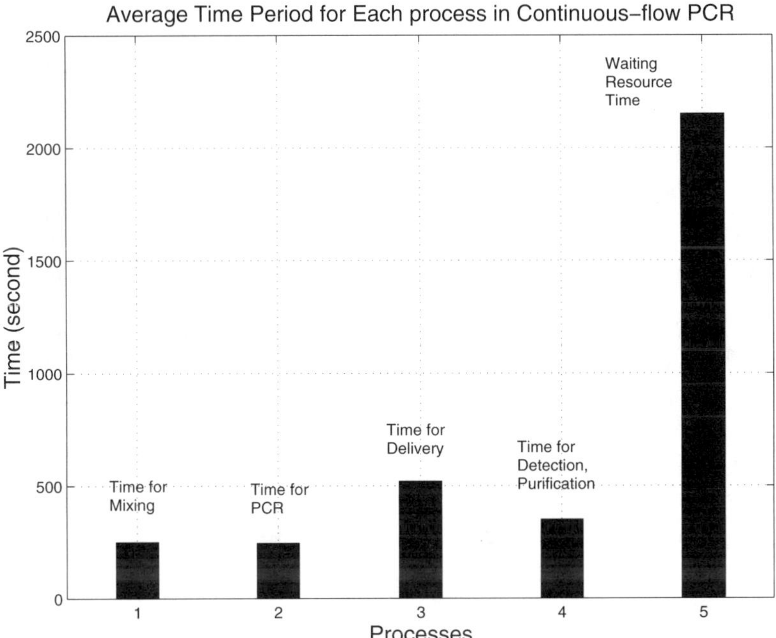


**FIGURE 6.26**
**Average time values for each process of the continuous-flow PCR system. The total time for each fluidic sample staying in the handling system consists of five periods. The DNA solution spends too much time waiting for system resources.**

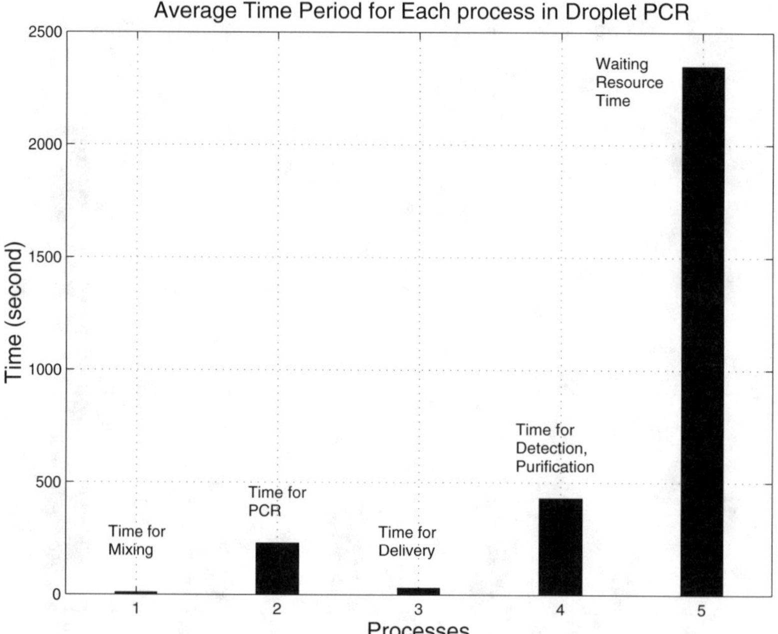


**FIGURE 6.27**
**Average time values for each process of the droplet-based PCR system. The total time for each fluidic sample staying in the handling system consists of five periods. The PCR block, and the detection purification block are two principal problems for system performance.**

extracting parallelism from a biochemical assay. As a case study, we apply our optimization method to the polymerase chain reaction, which is an important step in many lab-on-a-chip biochemical applications.

### 6.5.1 Introduction

Electrowetting-based actuation for microelectrofluidics systems has recently been proposed for optical switching[115], chemical analysis [105], and rotating yaw rate sensing [116]. Pollack *et al.* recently demonstrated that by varying the electrical potential along a linear array of electrodes, electrowetting techniques can be used to move liquid droplets along this line of electrodes [105]. By carefully controlling the electrical potential applied to the electrodes, fluid droplets can be moved as fast as $3cm/sec$ [105].

Electrowetting can also be used to move droplets in a two-dimensional electrode array. By controlling the voltage on the electrodes, fluid droplets can be moved freely to any location on a two dimensional plane [105]. Fluid droplets can also be confined to a fixed location and isolated from other droplets moving around it.

Using two-dimensional electrowetting arrays, many useful microfluidic operations can be performed, such as storing, mixing and droplet splitting. The store operation is performed by applying an insulating voltage around the droplet. This is analogous to a well. The insulating voltage prevents this droplet from mixing with other droplets around it. The mix operation is performed by routing two droplets to the same location, where they are merged into one droplet. Since the size of a droplet is kept small, effective mixing can be achieved by fluid diffusion after merging. Finally, the split operation is performed by creating opposite surface tension at the two ends of a fluid droplet and tearing it into two smaller droplets.

While two-dimensional electrowetting arrays are especially useful for biochemical analysis, system level design methodologies are required to harness this exciting new technology. In this section, we leverage electronic design automation techniques to develop the first system-level design methodology for reconfigurable MEFS-based lab-on-a-chip.

Reconfigurable computing systems based on field-programmable gate-arrays (FPGAs) are now commonplace [117]. However, the "programmability" of FPGAs is limited by the well-defined roles of interconnect and logic blocks. Interconnect cannot be used for storing information, and logic blocks cannot be used for routing. In contrast, the MEFS architecture that we are developing offers significantly more programmability. The grid points between electrodes can be used for storage, functional operations, as well as for transporting fluid droplets. Therefore, partitioning, resource allocation, and scheduling have emerged as major challenges for system-level MEFS design targeted at a set of biochemical applications.

We have developed the syntax and semantics of microfluidic operations such as MOVE, MIX, and SPLIT that can be used to describe biochemical processes such as Polymerase Chain Reaction (PCR) [118]. The various fluid samples represent the operands. Such a microfluidic program must then be mapped to the two-dimensional array that represents the datapath of a microfluidic computer. (A separate electronic control unit drives the electrodes.) The execution of microfluidic operations requires the availability of datapath resources (set of grid points) that can be appropriately configured. For example, the MIX operation requires that a set of grid points be properly configured to act as a mixer. The size of the two-dimensional array limits the number of concurrent operations of any type that can be carried out. Furthermore, functional dependencies between the operations in a microfluidic program also limit concurrency.

In Section 6.5.2, we describe the two-dimensional electrowetting array and introduce the concepts of virtual microfluidic components and partition maps. Section 6.5.3 presents the scheduling problem for biochemical analysis and describes an integer linear programming approach for scheduling under resource constraints in two-dimensional electrowetting arrays. Finally, Section 6.5.4 investigates the PCR reaction as a case study. The PCR reaction is an important step in LOC biochemical processing. Processing time must be minimized for a number of critical LOC applications such as the detection and identification of biochemical agents and health monitoring during surgery. Efficient scheduling techniques not only reduce processing time, but they also offer better resource utilization in two-dimensional electrowetting arrays.

### 6.5.2 Two-dimensional Electrowetting Array

A two-dimensional electrowetting array consists of a grid of electrodes on a two-dimensional plane (Figure 6.18). Fluid droplets are introduced to the device from the I/O ports on the boundary of the array. Droplets in the array have identical volumes. Hence, this type of device is also called a unit-flow device. It is desirable to maintain the unit-flow constraint since the rate of chemical and biomedical reactions grows exponentially with the growth of droplet volume [105].

Operations such as STORE, MOVE, MIX, and SPLIT are performed by controlling the electrical potential applied to the electrodes. It is easy to see that some of these operations violate the unit-flow assumption. For example, the fluid droplet size is likely to double as a result of a mixing operation. Therefore, we always perform a split operation after mixing to maintain the droplet volume.

In a continuous-flow MEFS system, mixing is performed using a micromixer. This is a specific device located at a fixed place in the microfluidic system. In unit-flow systems however, mixing operations can happen anywhere on the array, not necessarily at a specific location. If we define a mixer as the location where fluids mix, then a unit-flow mixer can be located at any arbitrary cell in the electrode grid. This property is referred to as reconfigurability, and it is in many ways similar to the reconfigurability provided by FPGAs. However, as discussed previously, unit-flow devices allow a higher degree of reconfigurability than FPGAs. Storage cells, mixers and splitters can be created, removed, and relocated at runtime. This allows us to create extremely flexible and efficient biochemical analysis systems.

An abstract model of the unit flow system with a two-dimensional grid of electrodes is shown in Figure 6.18. A ground plane is positioned above the electrode array at a spacing that is less than the diameter of the droplets. I/O ports are placed at the boundary of the system.

#### 6.5.2.1 Virtual Devices and Partition Maps

In the unit-flow environment, the routes that droplets travel and the rendezvous points of fluid droplets are programmed into a micro-controller that controls the voltages of electrodes. The storage and interconnect on the datapath are viewed as virtual devices by the controller.

A virtual device is defined to have three regions. The first is the functional region, where a particular function is performed. The second type of region is called the segregation region, which wraps around the functional region. This insulates the functional region from its environment. The outer-most region of the device is the inherited communication path. This provides a one-cell wide communication path for fluid droplet movement. Figure 6.28 shows a unit-flow storage cell. One droplet of a fluid sample is stored in each functional cell.

A partition map shows the time-varying positions of all the virtual devices inside the defined area. It is generated by the designer, and pre-loaded into the microcontroller, which then controls the electrode voltages according to partition map.

A partition map is similar to a virtual device, in that it is also a virtual map, and it only exists in the microcontroller specification. It is also dynamic in nature since it may change with time. Reconfiguration occurs when a new partition map is loaded into the controller. Figure 6.29 shows a partition map containing two storage cells, one input cell, and one mixer. (The labels A, B,..., J, K will be explained later.) The inherited communication paths of adjacent devices are combined to form a single channel in the electrode array. This channel is used for fluid droplet transfer, and is called a communication path. It forms the main network for fluid movement. Researchers have recently shown that it is possible to move the fluid droplets at a speed of 20 grids/second along this communication path [2]. The actual route along which a droplet moves is pre-determined and loaded into controller. If the routes of several consecutive droplets do not overlap, they are called compatible routes. Movements along compatible routes can be performed in parallel. If the routes are not compatible, the corresponding droplet movements must be performed sequentially.

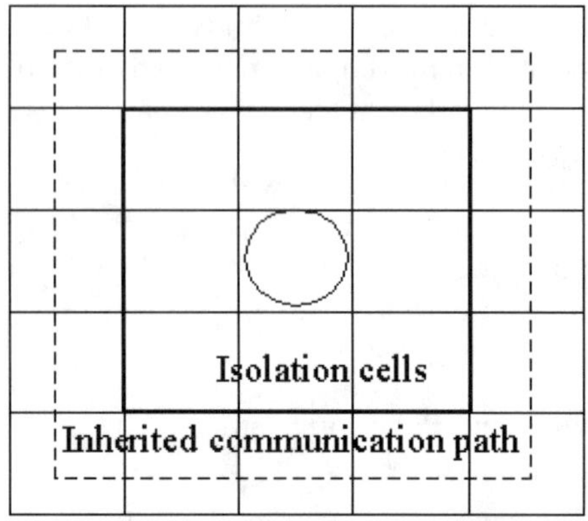


**FIGURE 6.28**
**A unit-flow storage device.**

We define the following operations that can be performed by virtual devices on a partition map.

- MIX mixer-name, where mixer-name is a reference to a particular mixer in the partition map.
- SPLIT mix-name, where mixer-name is a reference to a particular mixer in the partition map.
- INPUT port-name, fluid name, where port-name is a reference to a port in the partition map.
- MOVE source-name, destine-name, route-name, where route-name is a reference to a pre-defined path.
- PATH route-name, P1-P2-...-Pn, defines a path for droplet movement.

We next present a scheduling method for minimizing the processing time for fluid samples. We determine an optimal sequence of fluidic operations to minimize completion time under resource constraints (availability of virtual devices) and dependencies between operations.

In contrast to droplet movement, fluidic operations such as MIX and SPLIT are slow processes. The mixing by diffusion at the nanometer level takes about 1 minute for completion. During the same time period, a droplet can move along 1800 grid points. Therefore, we ignore droplet movement time for operation scheduling.

In order to schedule microfluidic operations such as MIX and SPLIT, we divide the time span between two consecutive reconfigurations into equal length time slots. The length of a time slot equals the greatest common divisor of all the operations. For example, if a MIX operation takes 3 minutes and a SPLIT operation takes 2 minutes, then the time slot is set to 1 minute. In this case, the MIX operation will take 3 slots, and the SPLIT operation will take 2 slots. In this way, we digitize the continuous fluid operation and the controller starts or completes an operation at the end of each time slot.

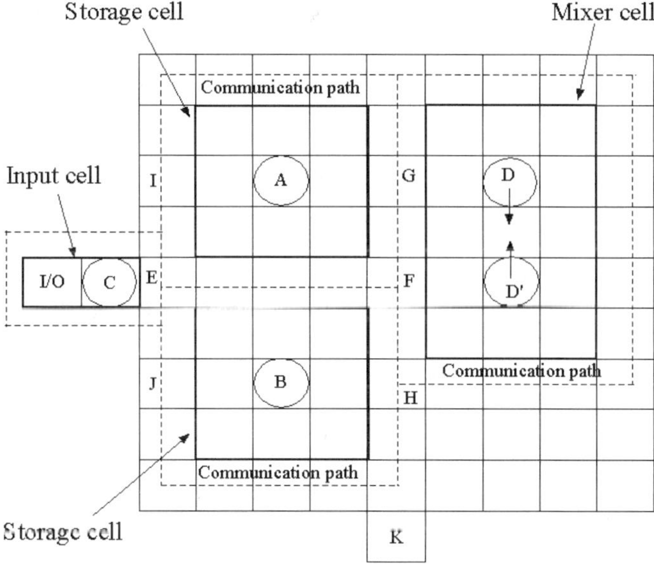


**FIGURE 6.29**
**Partition map with two storage units, one input cell, and one mixer.**

### 6.5.3 Schedule Optimization

The order of execution of microfluidic operations must be determined after carefully considering the dependencies between the operations and the availability of resources. While dependencies are imposed by the biochemical application, the resource constraints are imposed by the size of the two-dimensional electrowetting array and the availability of virtual devices. In this section, we use the dataflow graph model of high-level synthesis [119] to represent the scheduling problem and solve it using integer linear programming (ILP). The motivation for using ILP lies in the fact that it is a well-understood optimization method and we can leverage a number of public domain solvers [120].

First, each step of a biochemical process is represented using either a single microfluidic operation or a series of basic microfluidic operations. Each such instance of an operation forms a node in the dataflow graph. A directed edge from node $u$ to node $v$ indicates a dependency between the operations corresponding to $u$ and $v$, i.e. the operation corresponding to $u$ must be carried out before the operation corresponding to $v$. The goal of the scheduling problem is to determine the start times (time slots) of each operation so that the total completion time is minimized.

Let $x_{i,j}$ be a binary variable defined as follows:

$$x_{i,j} = \begin{cases} 1, & if\ operation\ i\ starts\ at\ time\ slot\ j \\ 0, & otherwise \end{cases}$$

where $1 \leq i \leq N$, the number of operations (nodes in the dataflow graph), and $1 \leq j \leq M$, the maximum possible index for a time slot. Note that $M$ can be trivially obtained by adding up the number of time slots required for all the operations. Note also that since each operation is scheduled exactly once, $\sum_{j=1}^{M} x_{i,j} = 1, \;\; 1 \leq i \leq N$.

The starting time $S_i$ for operation $i$ can now be expressed in terms of the set of variables $x_{i1}, x_{i2}, ..., x_{im}$. Assuming that each time slot is of length 1 unit, we get $S_i = \sum_{j=1}^{M} jx_{ij}$.

Each operation $i$ has an associated execution time $d_i$. If there exists a dependency edge between operation $i$ and operation $j$, then $s_j \geq s_i + d_i$. Such dependencies generally arise from the fluid samples that are used in each step of the biochemical reaction. These fluid samples are similar to variables in traditional architectural synthesis.

Finally, we add resource constraints to the ILP model. Let $a_k$ be an upper bound on the number of operations of type $k$. We now have the following set of constraints for each $k$: $\sum_{i \in T(k)} \sum_{j=l-d_i+1}^{l} x)ij \leq a_k, 1 \leq l \leq M$.

The objective of this optimization problem is to minimize the completion time of the last operation, i.e. **minimize** $max_i\{\sum_{j=1}^{M} jx_{ij} + d_i\}$. This can be linearized as: **minimize** $C$ subject to $C \geq \sum_{j=1}^{M}(jx_{ij}) + d_i, 1 \leq i \leq N$.

The ILP model can be easily solved using public-domain solvers. In our work, we used the lpsolve package from Eindhoven University of Technology in Netherlands [120].

### 6.5.4 Droplet-based PCR Systems

In this section, we present a case study for operation scheduling using the PCR reaction. The PCR reaction includes three basic steps. The first is the input section. In this part, a number of fluid samples are input into the system. Next, these samples are combined using a pre-determined set of *MIX* operations. Note that these are implemented by interleaving *MOVE*, *MIX*, and *SPLIT* operations. Finally, the sample mixture is sent off-chip for a series of heating steps.

The input samples for PCR include *Tris-HCl (pH 8.3), KCl, gelatin, bovine serum albumin, beosynucleotide triphosphate, a primer, AmpliTaq DNA polymerase*, and *lambdaDNA*. The PCR procedure consists of the following series of steps:

1. Introduce *Tris-HCl (pH 8.3)* to storage.
2. Introduce *KCl.*
3. Wait until *KCL* mixes with *Tris-HCl.*
4. Introduce *gelatin.*
5. Wait until it mixes.
6. Introduce *bovine serum albumin.*
7. Wait until it mixes.
8. Introduce *beosynucleotide triphosphate.*
9. Wait until it mixes.
10. Introduce *primer* to the storage.
11. Wait until it is well mixed with the mixture.
12. Introduce *AmpliTaq DNA polymerase.*
13. Wait until it is well mixed with the mixture.
14. Introduce *lambdaDNA* to the storage.
15. Wait until it is well mixed with the mixture.

#### 6.5.4.1 System configuration

The first example system we use is shown in Figure 6.29. The system can perform moving, mixing and splitting for the PCR reaction. It consists of 9-by-9 array of grid cells. A dedicated I/O port is located at the edge of the system. We assume that the mixing of two fluid droplets takes 2 minutes, while the input operation takes 0.4 minutes. Since the mix operation is always followed by a split operation, the latter is not explicitly considered here. Instead, we assume that the time for a split is included in the time for a mix operation. The speed of fluid movement is assumed to be 20 grid cells per minute.

The partition map for this example is given by Figure 6.29. In addition to the partition map, the droplet route plan and schedule of operations (to be determined next) must be loaded into the controller.

#### 6.5.4.2 Optimal Scheduling

We now describe how an optimal schedule can be derived to minimize the processing time. First, we represent the PCR reaction as a series of basic steps. This corresponds to a specification outlined by a lab technician, and serves as a user program. The user program can either be a sequential enumeration of steps, or it can contain a limited amount of hand-extracted concurrency. We then generate the dataflow graph based on the functional dependencies between the operations (Figure 6.30). An optimized PCR reaction for the datapath of Figure 6.18 and the dataflow graph of Figure 6.30 is given below:

The optimized PCR program of Table 6.10 was easy to derive since there is only one mixer in the system. The total processing time using this schedule is 14.8 minutes. We next show how the processing time can be decreased further and an optimal schedule derived using ILP.

Consider the partition map shown in Figure 6.31 with two mixers. This allows greater parallelism and demonstrates the advantage is using ILP to minimize the processing time. The following discussion presents the ILP model for this example in more detail.

The PCR program contains a total of 15 *INPUT* and *MIX* operations. From Table 6.10, we note that an upper bound on the processing time is 15 minutes. Each time slot is of length 0.4 minutes (the assumed time for an *INPUT* operation); hence an upper bound on the number of time slots is 37. There are 37 lots needed for the schedule. To build the ILP model for this partition map, we define a set of decision variables as discussed in Section 6.5.3. Thus our ILP model uses $x_{1,j} ... x_{15,j}$ as the decision variables, where $j = 1, 2, ..., 37$. The start time of each operation can be expressed as follows:

$$
\begin{aligned}
S_1 &= x_{1,2} + 2x_{1,3} + ... + 29x_{1,37} \\
S_2 &= x_{2,2} + 2x_{2,3} + ... + 29x_{2,37} \\
&... \\
S_15 &= x_{15,2} + 2x_{15,3} + ... + 29x_{15,37}
\end{aligned}
$$

**Table 6.10** Optimized PCR Reaction Based on the Datapath of Figure 6.30

| Time (minutes) | Operations | Representation |
|---|---|---|
| | Definition section<br>Path path1, C-E-F-G-D<br>Path path2, C-E-F-H-D'<br>Path path3, C-E-I-A<br>Path path4, C-E-J-B<br>Path path5, D'-F-H-K<br>Path path6, A-G-F-D'<br>Path path7, B-H-F-D' | Definition |
| 0 | Load partition map<br>*INPUT* Tris-HCl | <br>I1 |
| 0.4 | *MOVE* C, D, path1<br>*INPUT* KCl | <br>I2 |
| 0.8 | *MOVE* C, D', path2<br>*INPUT* gelatin<br>*MIX* D and D' | <br>I3<br>M1 |
| 1.2 | *MOVE* C, A, path3<br>*INPUT* bovine serum albumin | <br>I4 |
| 1.6 | *MOVE* C, B, path4 | |
| 2.8 | *MOVE* D', K , path5<br>*MOVE* A, D', path6<br>*INPUT* beosynucleotide triphosphate | <br><br>I5 |
| | *MIX* D and D' | M2 |
| 3.2 | *MOVE* C, A, path3 | |
| 4.8 | *MOVE* D', K, path5<br>*MOVE* A, D', path6<br>*INPUT* primer<br>*MIX* D and D' | <br><br>I6<br>M3 |
| 5.2 | *MOVE* C, A, path3 | |
| 6.8 | *MOVE* D', K, path5<br>*MOVE* A, D', path6<br>*INPUT* AmpliTaq DNA polymerase<br>*MIX* D and D' | <br><br>I7<br>M4 |
| 7.2 | *MOVE* C, A, path3 | |
| 8.8 | *MOVE* D', K, path5<br>*MOVE* A, D', path6<br>*INPUT* $\lambda$ DNA<br>*MIX* D and D' | <br><br>I8<br>M5 |
| 9.2 | *MOVE* Move C, A, path3 | |
| 10.8 | *MOVE* D', K, path5<br>*MOVE* A, D', path6<br>*MIX* D and D' | <br><br>M6 |
| 12.8 | *MOVE* D', K, path5<br>*MOVE* B, D', path7<br>*MIX* D and D' | <br><br>M7 |
| 14.8 | *MOVE* D', K, path5 | |

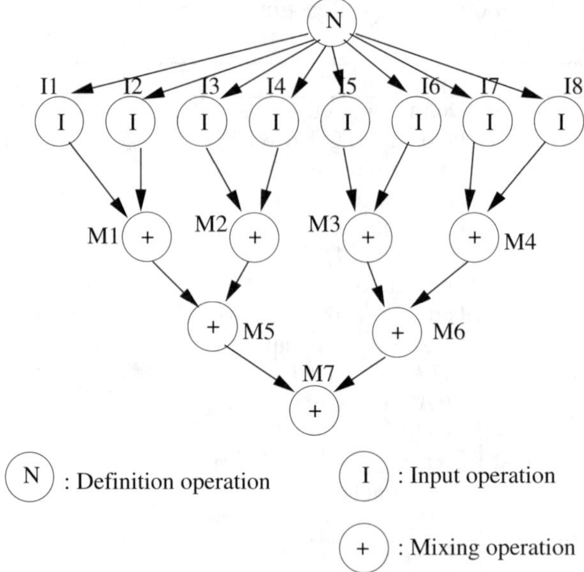


**FIGURE 6.30**
**Dataflow graph with input and mix operations.**

The dependency between instructions can be denoted using the following set of inequalities:

$$\begin{aligned} S_9 &\geq S_1, S_2 \\ S_10 &\geq S_3, S_4 \\ S_11 &\geq S_5, S_6 \\ &\dots \\ S_15 &= S_13, S_14 \end{aligned}$$

Finally, the resource constraints can be represented as:

$$\begin{aligned} x_{1,1} + x_{2,1} + \dots + x_{15,1} &\leq 2 \\ x_{1,2} + x_{2,2} + \dots + x_{15,2} &\leq 2 \\ &\dots \\ x_{1,37} + x_{2,37} + \dots + x_{15,37} &\leq 2 \end{aligned}$$

We solved this ILP model using *lpsolve*. It took 10 minutes of CPU time on a Sun Ultra Sparc with a 333 MHz processor and 128 MB of RAM. The optimum processing time is 9.6 minutes, $50\%$ faster than the PCR program of Table 6.10. The optimized schedule is given below:

This can be represented using the annotated dataflow graph shown in Figure 6.32.

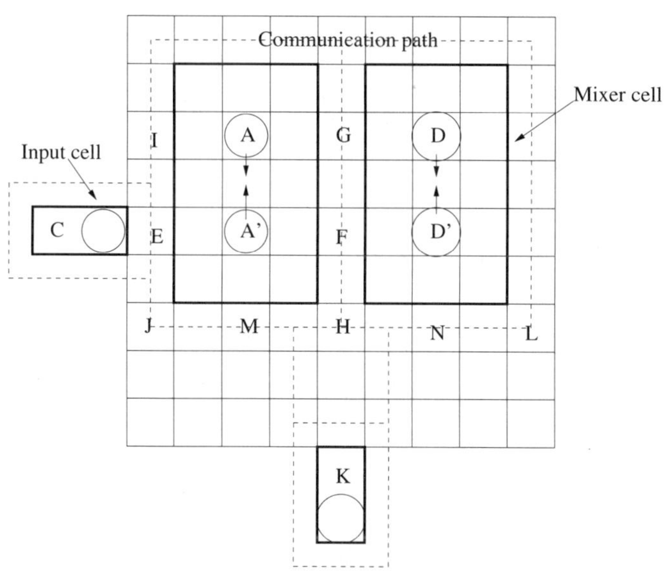


**FIGURE 6.31**
**Partition map with two mixers for PCR reaction.**

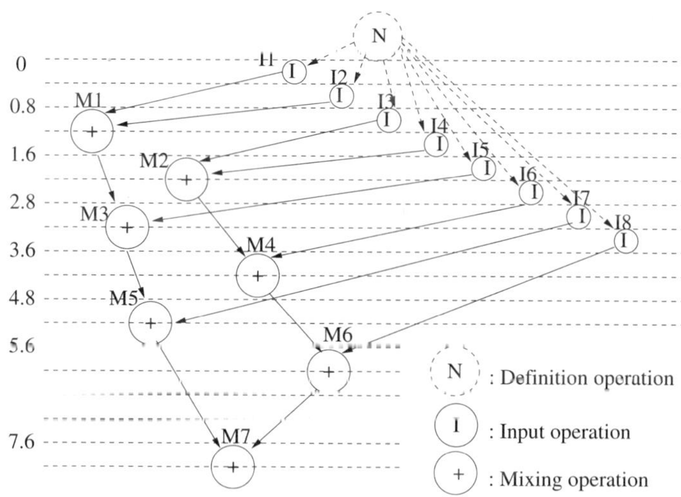


**FIGURE 6.32**
**Dataflow graph showing an optimized schedule for 2-mixer partition map.**

**Table 6.11** Optimized PCR Reaction

| Time (minutes) | Operations | Representation |
|---|---|---|
| | Definition section<br>Path path1, C-E-F-G-D<br>Path path2, C-E-F-H-D'<br>Path path3, C-E-I-A<br>Path path4, C-E-A'<br>Path path5, D'-F-H-K<br>Path path6, A-F-H-K<br>Path path7, A-G-F-D' | Definition |
| 0 | Load partition map<br>*INPUT* Tris-HCl | <br>I1 |
| 0.4 | *MOVE* C, D, path1<br>*INPUT* KCl | <br>I2 |
| 0.8 | *MOVE* C, D', path2<br>*INPUT* gelatin<br>*MIX* D and D' | <br>I3<br>M1 |
| 1.2 | *MOVE* C, A, path3<br>*INPUT* bovine serum albumin | <br>I4 |
| 1.6 | *MOVE* C, A', path4<br>*INPUT* beosynucleotide triphosphate<br>*MIX* A and A' | I5<br>M2 |
| 2.8 | *MOVE* D', K , path5<br>*MOVE* A, D', path1<br>*MIX* D and D'<br>*INPUT* primer | <br><br>M3<br>I6 |
| 3.6 | *MOVE* A', K, path6<br>*MOVE* C, A', path4<br>*MIX* A and A'<br>*INPUT* AmpliTaq DNA polymerase | |
| 4.8 | *MOVE* D', K, path5<br>*MOVE* C, D', path1<br>*MIX* D and D'<br>*INPUT* λDNA | <br><br>M5<br>I8 |
| 5.6 | *MOVE* D', K, path5<br>*MOVE* C, D', path1<br>*MIX* D and D' | <br><br>M6 |
| 6.8 | *MOVE* A', K, path6 | |
| 7.6 | *MOVE* D', K, path5<br>*MOVE* A, D', path7<br>*MIX* D and D' | <br><br>M7 |
| 9.6 | *MOVE* D', K, path5 | |

Consider next the partition map shown in Figure 6.33 with four mixers. This allows even greater parallelism and decreases processing time further. In the ILP model, we reformulated the resource constraints as follows:

$$x_{1,1} + x_{2,1} + ... + x_{15,1} \quad leq\ 4$$
$$\cdots$$
$$x_{1,37} + x_{2,37} + ... + x_{15,37}\ leq\ 4$$

The partition map for this implementation is shown in Figure 6.33.

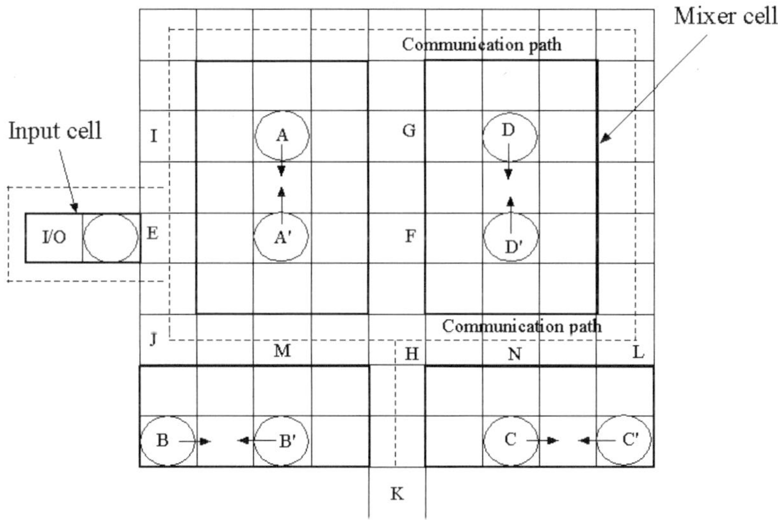


**FIGURE 6.33**
**Partition map with four mixers for PCR reaction.**

We solved this ILP and obtained the optimum processing time of 9.2 minutes, roughly $5.2\%$ faster than the 2-mixer version. Since the speed-up from two mixers to four mixers is insignificant, we conclude the maximum amount of parallelism in the PCR reaction has already been achieved.

We have presented a novel architectural design and optimization methodology for performing biochemical reactions using two-dimensional electrowetting arrays. We have defined a set of basic microfluidic operations and leveraged electronic design automation principles for system partitioning, resource allocation, and operation scheduling. While concurrency is desirable to minimize processing time, it is limited by the size of the two-dimensional array and functional dependencies between operations. We have used integer linear programming to minimize the processing time by automatically extracting parallelism from a biochemical assay. As a case study, we have applied our optimization method to the polymerase chain reaction. The proposed technique leverages known scheduling algorithms from electronic design automation for lab-on-a-chip (LOC) design, and it is expected to aid several LOC applications such as the rapid detection of biochemical agents, and reliable health monitoring during surgery.

# Chapter 7

## Conclusion

In this book, we presented a MEFS CAD closed-loop integration strategy along the line of microelectronics CAD. This strategy extends system design from the component level to the system level, and includes hierarchical modeling, hierarchical design environment, hierarchical performance evaluation, and hierarchical optimization.

In Chapter 2, we first defined basic variables needed to describe MEFS behavior at the low-level component layer. We have shown that lumped-element models with ODAEs are appropriate to describe MEFS dynamic behavior coupling with multiple energy domains. The equivalent circuit approach can be used for MEFS circuit-level device modeling and simulation, and its main drawbacks can be avoided by using modern hardware description languages. Next, we defined the description capacity requirements for MEFS system-level hierarchical modeling and performance evaluation. These requirements encompass system-level modeling, simulation, and statistical analysis.

In Chapter 3, we have developed a hierarchical integrated design environment with SystemC. First, we evaluated the suitability of several existing simulation languages for MEFS hierarchical design. These languages include VHDL/VHDL-AMS, SLAM, C/C++, Matlab, and SystemC. Then we showed that SystemC is a viable candidate for this purpose. Our design environment includes lower-level component modeling and simulation, as well as higher-level system modeling and simulation. The design environment consists of four different functional packages: system-level modeling package, circuit-level component modeling package, numerical simulation package, and optimization/verification package. These functional packages have been described in detail in the book.

We have presented a hierarchical modeling and simulation methodology in Chapter 4. This methodology combines high-level stochastic queuing networks with low-level nodal conservative differential equations. First, a more general microelectrofluidic system computational architecture for continuous-flow systems is introduced. Based on this generic reconfigurable architecture and the proposed modeling and simulation methodology, we have presented a complete system model and simulation results for a micro-chemical handling system based on the SystemC integrated design environment. By identifying the potential bottlenecks in the system, we have developed an improved MCHS architecture that reduces system processing time and increases resource availability. In addition, using the traffic variation method, we have investigated the system saturation process capacity. Moreover, we have analyzed the system-level performance sensitivity due to the variation of low-level component design parameters. Simulation results and performance analysis data have been presented for each case.

Due to growing design complexity, fabrication process variations, and the harsh operating environments of MEFS, there is a need for hierarchical design optimization to support all aspects of product development. To address this issue, we have demonstrated several MEFS design optimization methodologies in Chapter 5. First, we have proposed a statistical response analysis strategy. This strategy is useful to efficiently find an on-target design point that meets the performance goals, and it is also beneficial to find design parameters that need to be more carefully controlled during manufacturing. In addition, by leveraging Taguchi experimental design and statistical process control methods, we have demonstrated a robust design methodology. Moreover, on the analogy to a hardware/software co-design methodology, we have demonstrated a novel application flexibility design methodology. This method can be leveraged to extend the design and reuse it for additional applications. Several special MEMS and MEFS devices have been designed to illustrate these optimization algorithms.

Finally, based on the differences in flow control mechanisms, in Chapter 6, we have compared the performance of two types of microfluidic systems—continuous-flow systems and droplet-based systems. The comparison is based on a special microfluidic application—a polymerase chain reaction (PCR) system. The comparison includes system throughput, processing capacity, correction capability, and design complexity. We have demonstrated that the droplet-based microfluidic system provides higher performance, as well as lower design and integration complexity. Then, an architectural design and optimization methodology is presented for performing droplet-based biochemical reactions using two-dimensional electrowetting arrays.

In summary, this book is expected to pave the way for integrated top-down design of hierarchical MEFS. The framework described in this book will reduce design time and design cost, and increase system reliability.

The contents of this book open up a number of exciting directions for research. They are summaried below.

- Platform Transfer
  The physical implementation of MEFS reconfigurable architecture with the continuous-flow mechanism is very difficult, and several questions still remain unanswered. For instance, how can we evaluate the parasitic phenomena between microfluidic components? The droplet-based microfluidic system using electrowetting-based actuation mechanism is emerging as a promising approach to overcome these difficulties. This book provides an overview of an

integrated hierarchical design strategy for continuous-flow MEFS. The application platform to move from a continuous-flow system to a droplet-based system is left as a future research topic.

- Identification Applications
  It is important to identify a wider range of applications that can benefit from droplet-based technologies. Some of the applications that can be explored include polymerase chain reaction (PCR) [121], array population [21], and chemical analysis [105].

- Combination with ODAE Solvers
  Since SystemC does not provide an associated simulator, the designer is required not only to model the system behavior, but also to build the model solver. The problem of connecting ODAE models with the sophisticated ODAE solvers needs to be addressed. This linkage can relieve the system designer from the burden of simulator development.

- Multi-step Optimization
  Due to the complex design space, the proposed on-target design methodology may only lead to local optimal results. Therefore, alternative multi-step optimal on-target design methodologies need to be considered. The two-step heuristic/statistical on-target design method is especially promising. At first, the design solution space is divided into several sections. Then, a heuristic method such as generic algorithms (GA), simulation annealing (SA), or fuzzy logic is used to pre-select a solution area that contains the globally optimal solution. Next, using the statistical steepest ascent/descent method, the final globally optimal result can be obtained.

- Model Verification
  MEFS multi-parameter design requires model validation with respect to each design parameter. This means that the model is considered valid only if it is accurate within the scope of the variation of each design parameter. This is especially important since a model may be valid for one set of experimental conditions but invalid for another. In this book, the accuracy of these models are based on our initial assumptions. Therefore, finding a more efficient model verification method is necessary. This method must verify the accuracy of the model within the scope for each design parameter, from the nominal design point to the optimal design point. Such validation is critical for linking design optimization with device fabrication and field operation.

# Appendix A

## VHDL Queuing Model

List of some functional models.

- Mathematical Package MATH_REAL
  - Model name:
    **Package MATH_REAL (math_package.vhd)**
  - Functionality:
    VHDL declarations for mathematical package MATH_REAL which contains common real constants, common real functions, and real transcendental functions.
  - Interface definition:
    Related to each function parameter definition.
- Stochastic discrete-event model
  - Model name:
    **Queue (queue_model.vhd)**
  - Functionality:
    This model presents a queuing system with a single server using VHDL. The customer arrival rate is an exponential distribution with $\lambda = 60$. The server service rate is an Erlang distribution with $\lambda = 48$
  - Interface definition:
    No interface I/O port.
- Continuous system model
  - Model name:
    **Biology (biolo_model.vhd)**
  - Functionality:
    This is a program used to solve a biology system which is a continuous system. Here, we use the second-order Runge-Kutta method with variable step size to solve the differential equations.

$$dx/dt = rx(t) - Kx(t)y(t);$$

$$dy/dt = Kx(t)y(t) - Dy(t);$$

$x$ is the population of the host, and $y$ is the population of the parasites.

- Interface definition:
  No interface I/O port.

- Combined discrete-continuous system model
  - Model name:
    **Ingot (Ingot_model.vhd)**
  - Functionality:
    This model is used to solve a continuous and discrete combined system. The ingot arrival rate is an Uniform distribution between 10 to 15. The heating processing is based on differential equations.

$$dF/dt = 1.2 * (300 - F);$$

$$dP/dt = 0.3 * (F - p);$$

    where $F$ is the temperature of the oven, and $P$ is the temperature of the ingot.
  - Interface Definition:
    No interface I/O port.

# Appendix B

## Hierarchical Environment with SystemC

List of some functional models.

- Top-level model
  - Model name:
    **sc_main(int ac, char *av[]) (main.cpp)**
  - Functionality:
    This program is used to describe a biomedical system hierarchical modeling with SystemC language. The top level is used to define the communication protocol between several functional blocks.
  - Interface definition:
    No interface I/O port.
- Liquid sample creation model
  - Model name:
    **SC_MODULE(producer) (producer.h)**
  - Functionality:
    This model is used to generate a series of fluid samples
  - Interface Definition:

```
// port declaration
sc_outmaster<fluid_type > fluid_out_1; //
sc_outmaster<fluid_type > fluid_out_2; //
sc_in_clk clk;
```

- Liquid reservoir model
  - Model name:
    **SC_MODULE(reservoir) (reservoir.h)**

- Functionality:
  This model is used to define a liquid reservoir which is a first-in-first-out (FIFO) buffer.
- Interface definition:

```
// the connection to producer.
sc_inslave<fluid_type > fluid_write_1;
sc_inslave<fluid_type > fluid_write_2;

// the connection to mixer groups.
sc_outslave<fluid_type > fluid_read_1;
sc_outslave<fluid_type > fluid_read_2;

// two clock signal definition.
sc_in_clk wclk;
sc_in_clk rclk;
```

- Liquid mixer model
  - Model name:
    **SC_MODULE(mixing_unit) (ppu.h)**
  - Functionality:
    This model is used to describe the behavior of the mixer in a microfluidic system.
  - Interface definition:

```
 // input and output port definition.
// take fluid type 1
sc_inmaster<fluid_type > fluid_write_1;

// take fluid type 2
sc_inmaster<fluid_type > fluid_write_2;

// send fluid to processors
sc_outslave<fluid_type > fluid_out;

// used for check the status of the mixer.
sc_outslave<int > fulltest;

sc_in_clk  clk; // clock
```

- Processor group model
  - Model name:
    **SC_MODULE(processor) (processor.h)**
  - Functionality:
    This is the behavioral description of the processor block, we put all processors and bus sources in this module.
  - Interface definition:

```
// declare ports
// get the fluidic sample from the pre-processor.
sc_inmaster<fluid_type > fluid_write;

// send the fluidic sample out
sc_outmaster<fluid_type > fluid_out;

// check the pre-processor status
sc_inmaster<int > fulltest;

sc_in_clk  clk;    // clock.
sc_in_clk mathclk;
```

- Reactor model
  - Model name:
    **void processor::Reactor() (reactor.cpp)**
  - Functionality:
    This is the behavioral description of a reactor. The interface definition for several processors is the same, the difference between them is just about processing behavior definition.
  - Interface definition:

```
// declare ports
sc inmaster<fluid type > fluid write;
sc_outmaster<fluid_type > fluid_out; //
sc_inmaster<int > fulltest;
sc_in_clk  clk;
sc_in_clk mathclk;
```

- Terminal node model
  - Model name:
    **SC_MODULE (terminal) (terminal.h)**

- Functionality:
  This model is used to define a terminal node which can be used to record the statistical information.
- Interface definition:

```
// port definition
sc_inslave<fluid_type > fluid_write;
sc_in_clk wclk;
```

- Data process package
  - Model name:
    **Data_Function (general.cpp)**
  - Functionality:
    This is a general package to define several data processing functions.
  - Interface definition:
    Related to different functions.
- Mathematical package
  - Model name:
    **Math Package (math_package.cpp)**
  - Functionality:
    The is the mathematical package which contains common real constants, common real functions, and real transcendental functions.
  - Interface definition:
    Related to different functions.

# References


[1] Kymata Netherlands, "http://www.kymata.nl," .

[2] M. A. Burns *et al*, "An integrated nanoliter DNA analysis device," *Science*, vol. 282, pp. 484–487, 1998.

[3] Sandia National Laboratories, "http://www.sandia.gov," .

[4] R. Zengerle *et al.*, "A bidirectional silicon micropump," *Sensors and Actuators A*, vol. 50, pp. 81–86, 1995.

[5] S. McWhorter and S. A. Soper, "Conductivity detection of polymerase chain reaction products separated by micro-reversed-phase liquid chromatography," *Journal of Chromatography A*, vol. 883, no. 1-2, pp. 1–9, June 2000.

[6] T. Ohori, S. Shoji, K. Miura, and A. Yotsumoto, "Partly disposable three-way microvalve for a medical micro total analysis system (TAS)," *Sensors and Actuators A*, , no. 1, pp. 57–62, January 1998.

[7] M. U. Kopp, A J. de Mello, and Andreas Man, "Chemical amplification: Continuous-flow PCR on a chip," *Science*, vol. 280, pp. 1046–1048, May 1998.

[8] C. G. J. Schabmueller *et al.*, "Closed chamber PCR chips for DNA amplification," *Engineering Science and Education Journal*, vol. 9, no. 6, pp. 259–264, December 2000.

[9] Michael Pollack *et al.*, "MONARCH," http://www.ee.duke.edu/research/MONARCH/movies.html.

[10] W. Menz and A. Guber, "Microstructure technologies and their potential in medical applications," *Minimally Invasive Neurosurgury*, vol. 37, pp. 21–27, 1994.

[11] S. Shoji, "Microfabrication technologies and micro flow devices for chemical and bio-chemical micro flow systems," in *Proc. Int. Conf. Microprocesses and Nanotechnology, 1999*, 1999, pp. 72–73.

[12] R. G. Lerch *et al.*, "A programmable mixed-signal ASIC for data acquisition systems in medical implants," in *IEEE Int. Conf. Solid-State Circuits, Digest of Technical Papers. 41st ISSCC*, 1995, pp. 162–163, 357.

[13] B. Davies, "Robotics in minimally invasive surgery," in *IEE Colloquium Through the Keyhole: Microengineering in Minimally Invasive Surgery*, 1995, pp. 5/1–5/2.

[14] M. Fahndrich, B. Hochwind, and A. Zollner, "Fluid dynamics in micro dosing actuators," in *8th Int Conf. Solid-State Sensors and Actuators, and Eurosensors IX.*, 1995, vol. 2, pp. 295–298.

[15] K. Ikuta, T. Hasegawa, and T. Adachi, "SMA micro pumps and switching valves for biochemical IC family," in *Proc. Int. Symposium on Micromechatronics and Human Science*, 2000, pp. 169–174.

[16] J. Ulrich and R. Zengerle, "Static and dynamic flow simulation of a KOH-etched microvalve using the finite-element method," *Sensors and Actuators A*, vol. 53, pp. 379–385, 1996.

[17] J. Pfahler, J. harley, and H. Bau, "Liquid transport in micro and submicro channels," *Sensors and Actuators A*, vol. 22, 1990.

[18] P. Voigt, G. Schrag, and G. Wachutka, "Microfluidic system modeling using VHDL-AMS and circuit simulation," *Microelectronics Journal*, vol. 29, no. 11, pp. 791–797, November 1998.

[19] S. Senturia, "Microfabricated structures for the measurement of mechincal properties and adhesion of thin films," in *Proc. 4th Int. Conf. Solid-State Sensors and Actuators (Transducers '87)*, 1987, pp. 11–16.

[20] F. Maseeh, "A virtual prototype manufacturing software system for MEMS," in *Advanced Applications of Lasers in Materials Processing/Broadband Optical Networks/Smart Pixels/Optical MEMs and Their Applications. IEEE/LEOS 1996 Summer Topical Meetings*, 1996, p. 53.

[21] M. Tracey *et al.*, "A microfluidics-based instrument for cytomechanical studies of blood," in *Proc. 1st Int. Conf. Microtechnologies in Medicine and Biology*, 2000, pp. 62–67.

[22] J. R. Gilbert, "Integrating CAD tools for MEMS design," *Computer*, vol. 31, no. 4, pp. 99–101, April 1998.

[23] MEMSCAP, "http://www.memscap.com," .

[24] J. P. Hanna and R. G. Hillman, "A common basis for mixed-technology micro-system modeling," in *Proc. 2nd Int. Conf. Modeling and Simulation of Microsystems*, 1999, pp. 132–135.

[25] G. K. Fedder and Q. Jing, "A hierarchical circuit-level design methodology for microelectromechanical systems," *IEEE Transaction on Circuit and Systems II: Analog and Digital Signal Processing*, vol. 46, no. 10, pp. 1309–1315, October 1999.

[26] E. C. Russell, "Building simulation models with SIMSCRIPT II.5," Tech. Rep.

[27] A. Alan B. Pritsker, *Simulation with Visual SLAM and AweSim*, John Wiley and Sons, Inc., New York, NY, 1997.

[28] D. D. Gajski and R. H. Kuhn, "Guest editor's introduction: New VLSI tools," *IEEE Computer*, vol. 16, no. 12, pp. 4–8, 1983.

[29] P. I. Fierens and T. L. Fine, "Interval-valued probability modeling of internet traffic variables," in *Proc. IEEE Int. Symposium on Information Theory*, 2000, p. 80.

[30] G. Taguchi, *Taguchi on Robust Technology Development*, ASME, 1993.

[31] M. Montgomery, *Response Surface Methodology*, John Wiley and Sons, Inc., New York, NY, 1995.

[32] G. De Micheli and R. K. Gupta, "Hardware/software co-design," in *Proceeding of the IEEE*, March 1997, vol. 85, pp. 349–365.

[33] E. O. Doebelin, *System Dynamics: Modeling, Analysis, Simulation, Design*, Marcel Dekker, Inc., New York, NY, 1998.

[34] D. Rowell and D. N. Wormley, *System Dynamics: An Introduction*, Prentice Hall, Upper Saddle River, NJ, 1997.

[35] H. V. Vu and R. S. Esfandiari, *Dynamic Systems: Modeling and Analysis*, The McGraw-Hill Companies, Inc., New York, NY, 1997.

[36] A. Hargrove, "Modeling of acoustic noise attenuation by the use of composites," in *Proc. IEEE, Southeastcon '93*, 1993, p. 2p.

[37] P. W. Tuinenga, *SPICE : a guide to circuit simulation and analysis using PSpice*, Prentice Hall, Englewood Cliffs, NJ, 1995.

[38] Analogy Inc, Beaverton, OR, *Saber Designer Reference, Release 4.2*, 1997.

[39] H. A. C. Tilmans, "Equvialent circuit representation of electromechanical transducers II: Distributed-parameter systems," *Journal of Micromechanical and Microengineering*, vol. 7, pp. 285–309, 1997.

[40] T. Bourouina, "Modeling micropumps with electrical equivalent networks," *Journal of Micromechanics and Microengineering*, vol. 6, no. 4, pp. 398–404, 1996.

[41] B. Romanowicz, *Methodology for the Modeling and Simulation of Microsystems*, Kluwer Academic Publishers, New York, NY, 1998.

[42] W. L. Winston, *Operations Research : application and algorithms, 3rd Edition*, International Thomson Publication, New York, NY, 1994

[43] A. Dewey, *Analysis and Design of Digital Systems with VHDL*, PWS Publishing Company, New York, NY, 1997.

[44] B. Zeigler and K. Doohwan, "Design of high level modeling/high performance simulation environments," in *Proc. 10th IEEE Workshop on Parallel and Distributed Simulation PADS 96*, 1996, pp. 154–161.

[45] C. D. Pegden, *Introduction to Simulation Using SIMAN*, McGraw-Hill, Inc., New York, NY, 1995.

[46] T. Zhang *et al.*, "Performance analysis for microelectrofluidic system using hierarchical modeling and simulation," *IEEE Transactions on Circuit and System II : Analog and Digital Signal Processing*, vol. 48, pp. 482–491, May 2001.

[47] T. Zhang, A. Dewey, and R. B. Fair, "A hierarchical approach to stochastic discrete and continuous performance simulation using composable software components," *Microelectronics Journal*, vol. 31, no. 1, pp. 95–104, January 2000.

[48] C. Wohlin, C. Nyberg, and A. Larsson, "Reusable simulation models for performance analysis of intelligent networks," in *Proc. Conf. Intelligent Networks and New Technologies*, 1996, pp. 143–154.

[49] M. Orshansky, J. Chen, and H. Chenming, "A statistical performance simulation methodology for VLSI circuits," in *Proc. Design and Automation Conference*, 1998, pp. 402–407.

[50] B. Courtois *et al.*, "CAD tools and foundries to boost microsystems development," *Materials Science and Engineering B*, vol. 51, no. 1-3, pp. 242–253, February 1998.

[51] F. Gomes *et al.*, "A high performance logical process simulation class library in C++," in *Proc. IEEE Winter Simulation Conference 1995*, 1995, pp. 706–713.

[52] S. Habinc and P. Sinander, "Using VHDL for board level simulation," *IEEE Design & Test of Computers*, vol. 13, no. 3, pp. 66–78, 1996.

[53] A. Dewey *et al.*, "VHDL-AMS modeling considerations and styles for composite microsystems," in *Design Automation Conference*, 1999, Tutorial.

[54] F. Cao, "Microelectrofluidic systems (MEFS) component modeling and simulation," M.S. thesis, Duke University, 2000.

[55] B. Djafri and J. Benzakki, "OOVHDL: object oriented VHDL," in *Proc. VHDL International Users' Forum*, 1997, pp. 54–59.

[56] W. W. Dungan, R. H. Klenke, and J. H. Aylor, "Mixed-level modeling in VHDL using the watch-and-react interface," in *Proc. VHDL International Users' Forum*, 1997, pp. 25–32.

[57] J. Schoen, *Performance and Fault Modeling with VHDL*, Prentice-Hall, Inc., New York, NY, 1992.

[58] M. Srivastava, "Using VHDL for high-level, mixed-mode system simulation," *IEEE Design and Test of Computer*, vol. 9, no. 3, pp. 31–40, September 1992.

[59] B. Louis, *VHDL Programming with Advanced Topics*, John Wiley and Sons, Inc., New York, NY, 1993.

[60] J. Berge, *VHDL'92*, Kluwer Academic Publishers, New York, NY, 1993.

[61] G. Gordon, *System Simulation*, Prentice-Hall, New York, NY, 1978.

[62] B. Soong, "Object-oriented description of queuing systems for flow control studies," in *Proc. Int. Conf. IEEE Singapore Networks / Int. Conf. Information Engineering 1995. Theme: Electrotechnology 2000: Communications and Networks*, 1995, pp. 447–449.

[63] A. Law, *Simulation Modeling and Analysis*, McGraw-Hill, Inc., New York, NY, 1991.

[64] R. L. Burden, *Numerical Analysis*, Prindle, Weber and Schmidt, Boston, MA, 1985.

[65] Z. Dimic *et al.*, "VHDL-AleC++ co-simulation," in *Proc. 21st Int. Conf. Microelectronics (MIEL'97)*, Nis, Yugoslavia, 1997, vol. 2, pp. 14–17.

[66] G. Arnout, "SystemC standard," 2000, pp. 573–577.

[67] T. Barta, "A quantitative comparison of three simulation languages : SLAM, GPSS/H, SIMSCRIPT," *Comput. & Indus. Eng.*, vol. 9, no. 1, pp. 45–66, 1985.

[68] R. Pratap, *Getting started with MATLAB : a quick introduction for scientists and engineers*, Saunders College Pub., Fort Worth, Texas, 1996.

[69] D. Ramanathan, R. Roth, and R. Gupta, "Interfacing hardware and software using C++ class libraries," in *Proc. Int. Conf. Computer Design*, 2000, pp. 445–450.

[70] S. Y. Liao, "Towards a new standard for system-level design," in *Proc. 8th Int. Workshop on Hardware/Software Codesign*, 2000, pp. 2–6.

[71] H. J. Schlebusch, "SystemC based hardware synthesis becomes reality," in *Proc. 26th Int. Euromicro Conference*, 2000, vol. 1, p. 434.

[72] *SystemC Release Beta v1.1 Reference Manual*, http://www.systemc.org.

[73] Z. Mrcarica *et al.*, "Hierarchical modeling of microsystems in an object-oriented hardware description language," in *Proc. 21st Int. Conf. on Microelectronics (MIEL'97)*, 1997, pp. 14–17.

[74] D. H. van der Schoot *et al.*, "A silicon integrated miniature chemical analysis system," *Sensors and Actuators B*, vol. 6, pp. 57–60, 1992.

[75] T. Bourouina, "Design and simulation of an electrostatic micropump for drug-delivery application," *Journal of Micromechanics and Microengineering*, vol. 7, no. 3, pp. 186–188, 1997.

[76] R. B. Fair *et al.*, "MONARCH," http://www.ee.duke.edu/research/MONARCH/index.html.

[77] A. Dewey *et al.*, "Towards microfluidic system (MEFS) computing and architecture," in *Proc. 3rd Int. Conf. Modeling and Simulation of Microsystems (MSM2000)*, 2000, pp. 142–145.

[78] M. Kirihara and K. Taniguchi, "Monte Carlo simulation for single electron circuits," in *Proc. Asia and South Pacific Design Automation Conference*, 1997, pp. 333–337.

[79] Z. R. Yang, "Bootstrap, an alternative to Monte Carlo simulations," *Electronics Letters*, vol. 34, no. 12, pp. 1174–1175, 1998.

[80] R. Tibshirani, "A comparison of some error estimates for neural network models," *Neural Computation*, vol. 8, pp. 152–163, 1996.

[81] A. M. Zoubir, "The bootstrap and its application in signal processing," *IEEE Signal Processing Magazine*, vol. 11, no. 5, pp. 56–76, January 1998.

[82] W. C. Tang *et al.*, "Electrostatic comb drive of lateral polysilicon resonators," *Sensors and Actuators A*, vol. 21, pp. 328–331, 1990.

[83] C. T. Nguyen, "Electromechanical characterization of microresonators for circuit applications," M.S. thesis, UC Berkeley, 1991.

[84] R. T. Howe, "Resonant microsensors," in *Technical Digest, 4th Int. Conf. Solid-State Sensors and Actuators (Transducers'87)*, 1987, pp. 843–848.

[85] C. T. Nguyen, *Micromechanical Signal Processors*, Ph.D. thesis, UC Berkeley, 1994.

[86] T. C. Nguyen and R.T. Howe, "Quality factor control for micromechanical resonators," in *IEDM*, 1992, pp. 505–508.

[87] A. Dewey, T. Zhang, H. Ren, and J. Wu, "Parametric regression analysis of microelectromechanical resonator," in *Proc. Int. Behavior Modeling and Simulation Workshop (BMAS98)*, 1998.

[88] G. A. F. Seber, *Linear Regression Analysis*, John Wiley and Sons, New York, NY, 1977.

[89] J. Neter, *Applied Linear Statistical Models*, Irwin Publishing, 1996.

[90] S. M. Ross, *Introduction to Probability Models*, Academic Press, San Diego, CA, 1997.

[91] B. L. Raktoe, A. Hedayat, and W. T. Federer, *Factorial Designs*, John Wiley and Sons, New York, NY, 1981.

[92] A. S. Hedayat, *Orthogonal Arrays: Theory and Applications*, Spring-Verlag New York, Inc., New York, NY, 1999.

[93] R. G. Sargent, "Verification and validation of simulation models," in *Proc. the 1998 Winter Simulation Conference*, 1998, pp. 121–129.

[94] M. Courtoy, "Rapid prototyping for communications design validation," in *Southcon/96. Conference Record*, 1996, pp. 49–54.

[95] Schlesinger *et al.*, "Terminology for model credibility," *Simulation*, vol. 32, no. 3, pp. 103–104, 1979.

[96] M. Fujita, "Model checking: its basics and reality," in *Proc. Asia and South Pacific Design Automation Conference*, 1998, pp. 217–222.

[97] A. Dewey, H. Ren, and T. Zhang, "Behavior modeling of microelectromechanical systems (MEMS) with statistical performance variability reduction and sensitivity analysis," *IEEE Transactions on Circuit and Systems II: Analog and Digital Signal Processing*, vol. 47, no. 2, pp. 105–113, 2000.

[98] D. Young, "Application of statistical design and response surface methods to computer-aided VLSI device design II: Desirability functions and Taguchi methods," *IEEE Transactions on Computer Aided Design*, vol. 10, no. 1, pp. 103–115, 1991.

[99] J. W. Knutti, "Finding market for microstructures," in *Proc. SPIE Conf. Microfluidic Devices and Systems*, 1998, pp. 17–23.

[100] C. T.-C Nguyen, "Micromechanical resonators for oscillators and filters," in *Proc. 1995 IEEE International Ultrasonics Symposium*, 1995, pp. 489–499.

[101] J. C. Zhang and M. A. Styblinski, *Yield and Variability Optimization of Integrated Circuits*, Kluwer Academic Publishers, New York, NY, 1995.

[102] M. Gorges-Schleuter *et al.*, "An evolutionary algorithm for design optimization of microsystems," in *Proc. 4th Int. Conf. Parallel Problem Solving from Nature*, 1996, pp. 1022–1031.

[103] Analog Devices, *Monolithic Accelerometer With Signal Conditioning*, Norwood, MA, 1996.

[104] M. Washizu, "Electrostatic actuation of liquid droplets for micro-reactor applications," in *IEEE Industry Applications Society Annual Meeting*, 1997, pp. 1867–1873.

[105] M. Pollack, R. B. Fair, and A. Shenderov, "Electrowetting-based actuation of liquid droplets for microfluidic applications," *Applied Physical Letters*, vol. 77, no. 11, pp. 1725–1726, July 2000.

[106] L. Stryer, *Biochemistry*, W. H. Freeman and Company, New York, NY, fourth edition, 1995.

[107] PT Wire, "http://www.sisweb.com," Scientific Instrument Service, Ringoes, NJ, USA.

[108] S. Kar *et al.*, "Bipolar-pulsed technique solution conductivity," *Analytical Chemistry*, vol. 66, pp. 2537, 1994.

[109] H. Swerdlow *et al.*, "Fully automated DNA reaction and analysis in a fluidic capillary instrument," *Analytical Chemistry*, vol. 69, no. 5, pp. 848–855, 1997.

[110] S. Shoji *et al.*, "Smallest dead volume microvalves for integrated chemical analyzing systems," in *Int. Conf. Solid-State Sensors and Actuators, 1991*, 1991, pp. 1052–1055.

[111] T. S. J. Lammerink *et al.*, "Modular concept for fluid handling systems: A demonstrator micro analysis system," in *IEEE 9th Annual Int. Workshop on Micro Electro Mechanical Systems, MEMS '96*, 1996, pp. 389–394.

[112] A. Rasmussen and M. E. Zaghloul, "The design and fabrication of microfluidic flow sensors," in *Proc. Int. Symposium on Circuits and Systems*, 1999, vol. 5, pp. 136–139.

[113] C. G. J. Schabmueller *et al.*, "Design and fabrication of a microfluidic circuitboard," *Journal of Micromechanical and Microengineering*, vol. 9, pp. 176–179, 1999.

[114] N. Schwesinger, T. Frank, and H. Wurmus, "A modular microfluid system with an integrated micromixer," *Journal of Micromechanics and Microengineering*, vol. 6, pp. 99–102, March 1996.

[115] J. L. Jackel *et al.*, "Electrowetting switch for multimode optical fibers," *Applied Optics*, vol. 22, no. 11, pp. 1765–1770, 1999.

[116] R. Yates *et al.*, "A micromachined rotating yaw rate sensor," in *Micromachined Devices and components II, SPIE meeting*, 1996, pp. 161–168.

[117] S. M. Trimberger, *Field-programming Gate Array Technology*, Kluwer Academic Publishers, Norwell, MA, 1994.

[118] L. C. Waters *et al.*, "Multiple sample PCR amplification and eletrophoretic analysis on a microchip," *Analytical chemistry*, vol. 70, no. 24, December 1998.

[119] G. De Micheli, *Synthesis and Optimization of Digital Circuits*, McGraw-Hill, Inc., New York, NY, 1994.

[120] M. Berkelaar, *lpsolve 3.0*, University of Technology, Eindhoven, The Netherlands, ftp://ftp.ics.ele.tue.nl/pub/lp_solve.

[121] J. Ding, K. Chakrabarty, and R. B. Fair, "Reconfigurable microfluidic system architecture based on two-dimensional electrowetting arrays," in *Proc. Int. Conf. Modeling and Simulation of Microsystems*, 2001, pp. 181–185.

[122] G. Blankenstein and U. D. Larsen, "Modular concept of a laboratory on a chip for chemical and biochemical analysis," *Biosensors and Bioelectronics*, vol. 13, no. 3-4, pp. 427–438, March 1998.

[123] J. White *et al.*, "A multi-level newton method for static and fundamental frequency analysis of electromechanical systems," in *Proc. Int. Conf. Simulation of Semiconductor Processes and Devices*, 1997, pp. 125–128.

[124] G. C. Goodwin and R. L. Payne, *Dynamic System Identification: Experiment Design and Data Analysis*, Academic Press, New York, NY, 1977.

[125] R. Bagrodia and C. Shen, "MIDAS: Integrated design and simulation of distributed systems," *IEEE Transactions on Software Engineering*, vol. 17, no. 10, pp. 1042–1058, 1991.

[126] S. Buttgenbach, "Spotlights on recent developments in microsystem technology," in *European Design Automation Conference*, 1996, pp. 274–276.

[127] M. Carmona *et al.*, "Dynamic simulations of micropumps," *Journal of Micromechanics and Microengineering*, vol. 6, no. 1, pp. 128–130, March 1996.

[128] M. R. Chen, "VLSI process optimization using Taguchi method with multiple-criteria approach," in *3rd Int. Workshop on Statistical Metrology*, 1998, pp. 96–99.

[129] J. S. Go and Y. H. Cho, "Experimental evaluation of anodic bonding process using Taguchi method for maximum interfacial fracture toughness," in *Proc. Int. Workshop on Micro Electro Mechanical Systems(MEMS98)*, 1998, pp. 318–321.

[130] P. Dario *et al.*, "A fluid handling system for a chemical microanalyzer," *Journal of Micromechanics and Microengineering*, vol. 6, no. 1, pp. 95–98, March 1996.

[131] R. C. Degeneff *et al.*, "Nonlinear, lumped parameter transformer model reduction technique," *IEEE Trans. Power Delivery*, vol. 10, no. 2, pp. 862–868, April 1995.

[132] R. C. Degeneff *et al.*, "Linear, lumped parameter transformer model reduction technique," *IEEE Trans. Power Delivery*, vol. 10, no. 2, pp. 853–862, April 1995.

[133] J. D. Lee *et al.*, "A monolithic thermal inkjet printhead utilizing electrochemical etching and two-step electroplating techniques," in *Int. Electron Devices Meeting*, 1995, pp. 601–604.

[134] R. S. Esfandiari, *Dynamic Systems Modeling and Analysis*, McGraw-Hill, Inc., New York, NY, 1995.

[135] D. Gazal, A. Benzekri, and Y. Raynaud, "Integration of performance simulation in the systems development life cycle," in *Proc. 11th Int. Conf. Applied Informatics, IASTED*, 1995, pp. 103–106.

[136] R. Zengerle and M. Ritcher, "Simulation of microfluid systems," *Journal of Micromechanics and Microengineering*, vol. 4, no. 4, pp. 192–204, December 1994.

[137] T. Gerlach and H. Wurmus, "Working principle and performance of the dynamic micropump," *Sensors and Actuators A*, vol. 50, no. 1-2, pp. 135–140, August 1995.

[138] *ANSYS user's manual*, Houston, PA, revision 5.2 edition, 1995.

[139] P. Gravesen, J. Branebjerg, and J. Jensen, "Microfluidics - a review," *Journal of Micromechanics and Microengineering*, vol. 3, no. 4, pp. 168–182, December 1993.

[140] J. Carroll, "FET statistical modeling using parameter orthogonalization," *IEEE Trans. Microwave Theory and Techniques*, vol. 44, no. 1, pp. 47–55, January 1996.

[141] B. Kanchana and V. V. Sarma, "Software quality enhancement through software process optimization using Taguchi methods," in *Proc. IEEE Conference and Workshop on Engineering of Computer-Based Systems, ECBS '99*, 1999, pp. 188–193.

[142] J. M. Karam *et al.*, "CAD and foundries for microsystems," in *Proc. Design Automation Conference*, 1997, pp. 674–679.

[143] K. Nellayappan, "SEAMS: A mixed-signal simulation environment for VHDL-AMS with emphasis on run-time elaboration and analog design partitioning," M.S. thesis, University of Cincinnati, 1995.

[144] D. Kincaid, *Numerical Analysis*, Brooks/Cole, 1996.

[145] K. D. Mueller-Glaser, "CAD of microsystems - a challenge for systems engineering," in *European Design Automation Conference (EURO-DAC)*, 1996, pp. 280–281.

[146] G. M. Koppelman, "OYSTER, a three-dimensional structural simulator for microelectromechanical design," *Sensors and Actuators, A*, vol. 20, pp. 179–185, 1989.

[147] W. Kuehnel and S. Sherman, "A surface micromachined silicon accelerometer with on-chip detection circuitry," *Sensors and Actuators A*, vol. 45, pp. 7–16, 1994.

[148] H. Leuthold and F. Rudolf, "An ASIC for high-resolution capacitive microaccelerometers," *Sensors and Actuators A*, vol. 21, no. 1-3, pp. 39–47, October 1990.

[149] Analogy Inc, Beaverton, OR, *MAST Reference Manual, Release 4.2*.

[150] T. Lammerink, M. Elwenspoek, and J. Fluitman, "Integrated micro liquid dosing system," in *Proc. IEEE Micro Electro Mechanical Systems - An In-*

*vestigation of Micro Structures, Sensors, Actuators, Machines, and Robotic Systems*, 1993, pp. 254–259.

[151] L. Geppert, "Electronic design automation," *IEEE Spectrum*, pp. 70–74, January 2000.

[152] S. Meinzer *et al.*, "Simulation and design optimization of microsystems based standard simulators and adaptive search techniques," in *Proc. the EURO DAC '96 - European Design Automation Conference*, September 1996, pp. 322–327.

[153] T. Mukherjee *et al.*, "Structured design of microelectromechanical systems," in *Proc. Design Automatic Conference 97 (DAC97)*, 1997, pp. 680–685.

[154] T. Mukherjee and G. K. Fedder, "Design methodology for mixed-domain systems-on-a-chip MEMS design," in *Proc. IEEE Computer Society Workshop on System Level Design*, 1998, pp. 96–101.

[155] O. Nagler *et al.*, "Efficient design and optimization of MEMS by integrating commercial simulation tools," in *Proc. Int. Conf. Conf. Solid-State Sensors and Actuator, TRANSDUCERS' 97*, 1997, pp. 1055–1058.

[156] B. Romanowicz *et al.*, "Modeling and simulation of electromechanical transducers in microsystems using an analog hardware description language," in *Proc. Conf. European Design and Test, 1997, ED&TC 97*, 1997, pp. 119–123.

[157] S. Senturia, "CAD for microelectromechanical systems," in *Proc. 8th Int. Conf. Solid-State Sensors and Actuators, and Eurosensors IX*, 1995, vol. 2, pp. 5–8.

[158] S. Senturia, "CAD challenges for microsensors, microactuators, and microsystems," *Proceedings of the IEEE*, vol. 86, no. 8, pp. 1611–1626, August 1998.

[159] D. L. Smith, *Introduction to Dynamic Systems Modeling for Design*, Prentice Hall, Englewood Cliffs, NJ, 1994.

[160] A. Takaki *et al.*, "Protection of tombstone problems for small chip devices," in *Proc. 49th. Conf. Electronic Components and Technology*, 1999, pp. 1036–1041.

[161] W. C. Tang, *Electrostatic Comb Drive for Resonant Sensor and Actuator Applications*, Ph.D. thesis, UC Berkeley, 1989.

[162] W. C. Tang *et al.*, "Electrostatic comb drive of lateral polysilicon resonators," *Sensors and Actuators A*, vol. 21, pp. 328–331, 1990.

[163] H. A. C. Tilmans, "Equivalent circuit representation of electromechanical transducers I: Lumped-parameter systems," *Journal of Micromechanical and Microengineering*, vol. 6, pp. 157–176, 1996.

[164] S. Timoshenko, D. Young, and W. Weaver, *Vibration Problems in Engineering*, John Wiley and Sons, New York, NY, 1974.

[165] P. Voigt, G. Schrag, and G. Wachutka, "Electrofluidic full-system modeling of a flap valve micropump based on Kirchoffian network theory," *Sensors and Actuators A*, vol. 66, pp. 9–14, 1998.

[166] J. C. Zhang and M. A. Styblinski, *Yield and Variability Optimization of Integrated Circuits*, Kluwer Academic Publishers, 1995.

# Index